TOWARDS DRUGS OF THE FUTURE

Solvay Pharmaceuticals Conferences

Series Editors

Werner Cautreels, Claus Steinborn and Lechoslaw Turski

Volume 9

Previously published in this series

Vol. 8	J.L. Junien and B. Staels (Eds.), Nuclear Receptors as Molecular Targets for Cardiometabolic and Central Nervous System Diseases
Vol. 7	B. Maisch and R. Oelze (Eds.), Cardiovascular Benefits of Omega-3 Polyunsaturated Fatty Acids
Vol. 6	B. Testa and L. Turski (Eds.), Virtual ADMET Assessment in Target Selection and Maturation
Vol. 5	C.G. Kruse, H.Y. Meltzer, C. Sennef and S.V. van de Witte (Eds.), Thinking About Cognition: Concepts, Targets and Therapeutics
Vol. 4	J. Antel, N. Finer, D. Heal and G. Krause (Eds.), Obesity and Metabolic Disorders
Vol. 3	G. Krause, J.R. Malagelada and U. Preuschoff (Eds.), Functional Disorders of the Gastrointestinal Tract
Vol. 2	J.G. Papp, M. Straub and D. Ziegler (Eds.), Atrial Fibrillation: New Therapeutic Concepts
Vol. 1	E. Ronken and G.J.M. van Scharrenburg (Eds.), Parkinson's Disease

ISSN 1566-7685

Towards Drugs of the Future

Key Issues in Lead Finding and Lead Optimization

Edited by

C.G. Kruse

Solvay Pharmaceuticals, Weesp, The Netherlands

and

H. Timmerman

Professor emeritus Pharmacochemistry,
VU University Amsterdam, The Netherlands

IOS Press

Amsterdam • Berlin • Oxford • Tokyo • Washington, DC

ISBN 978-1-58603-949-3
Library of Congress Control Number: 2008940898

Publisher
IOS Press
Nieuwe Hemweg 6B
1013 BG Amsterdam
Netherlands
fax: +31 20 687 0019
e-mail: order@iospress.nl

Distributor in the UK and Ireland
Gazelle Books Services Ltd.
White Cross Mills
Hightown
Lancaster LA1 4XS
United Kingdom
fax: +44 1524 63232
e-mail: sales@gazellebooks.co.uk

Distributor in the USA and Canada
IOS Press, Inc.
4502 Rachael Manor Drive
Fairfax, VA 22032
USA
fax: +1 703 323 3668
e-mail: iosbooks@iospress.com

PRINTED IN THE NETHERLANDS

Towards Drugs of the Future
C.G. Kruse and H. Timmerman (Eds.)
IOS Press, 2008

Preface

"The Solvay Pharmaceuticals Conferences: where industry meets academia in a search for novel therapies"

The Big Picture of Pharmaceutical Industry is Changing

To capture growth opportunities of the future the pharmaceutical industry must change the way it operates [1]. In a new regulatory and stakeholder environment the current business model is economically unsustainable and incapable to deliver expected output [1,2]. The number of novel medications in the global industry pipeline is insufficient to suffice demand of the markets, the financial performance of several companies is declining, sales and marketing costs are rising, regulatory constraints are increasingly challenging and overall reputation of industry is tarnishing. Unmet medical needs must be better addressed and tangible benefits need to be exposed in order to be rewarded by healthcare payers [3]. In addition, the focus is shifting from simple treatment of the ill to preventive measures and health management. To achieve this fundamental change industry will shift investment again towards scientific approaches, in life sciences and towards (pharmaco)economics, and away from conventional sales and marketing.

A new model may be described as follows: The linear phase development process will be replaced by in-life testing and life licensing, the initial clinical studies will become smaller but will be more focused and the results will be shared with regulators when testing. Safety and efficacy data will be globally shared by regulatory agencies which may lead to one global regulatory system and a larger safety data base. Integrated packages of medicines and services will dominate pharmaceuticals market and soon disease management may create more value than medications themselves. The distribution channels of medicines will be reinforced by enabling technologies allowing for e-dispensing of medicines directly to patients from online pharmacies [3]. Such global challenges require pharmaceutical industry to rethink and adopt the current business model in order to contribute value to the healthcare systems of the future [4].

To retain key position of chemistry in the future pharmaceutical industry the way chemists work must change. During the last two decades paramount investments in mass screening technologies and in combinatorial chemistry were made, knowledge in synthetic chemistry expanded, computer-assisted drug design matured but output of medicinal chemistry, measured as successful launches of novel medicines, declined indicating that novel technologies are used in old paradigms [1,5].

The challenge is to ensure that new chemistry is driving progress of pharmaceutical industry by means of novel technologies in which system biology approach or target approach is computer simulated, testing is performed *in silico* on rat or human digital models and synthesis of molecules is restricted to intelligent synthesis of treatments based on simulations. Implementation of these technologies will reduce costs and shorten timelines because only intelligent treatments will be tested *in vitro* and *in vivo* and portfolio decisions will be based on *in silico* tests while pharmacology and toxicology tests on animals will only be performed on candidates selected as intelligent treatments [2,6].

The Ninth Solvay Pharmaceuticals Conference on Medicinal Chemistry held in Garmisch-Partenkirchen (Germany) September 26–28, 2007 addressed newest development in synthetic and medicinal chemistry with the emphasis on the challenge of the future. The scientific focus of this conference centered on the involvement of emerging technologies in chemistry on the speed of drug discovery and on the aspects of drug finding including intelligent biology and rational drug design.

W. Cautreels
C. Steinborn
L. Turski

References

[1] J. Cacciotti and B. Shew. Pharma's next top model: slimmer business models. *Pharm. Exec.* **3** (2006) 28–30.
[2] EFPIA. The Pharmaceutical Industry in Figures (2008).
[3] S. Arlington. A vision of 2020. *Pharm. Exec Europe* **10** (2007) 26–28.
[4] I. Massey. An alternative drug development model. *Pharm. Exec. Europe* **9** (2008) 24–25.
[5] L. Turski. New business models required for today's drug development. *Conceptuur* **44** (2005) 6–7.
[6] L. Turski. A passion for progress. *GIT Lab. J.* **8** (2004) 37.

List of Contributors

Afshar, M.
Ariana Pharma, BioTop, Institut Pasteur, 28 rue Dr Roux, 75724 Paris CEDEX 15, France

Borchardt, Ronald T.
Department of Pharmaceutical Chemistry, 104 McCollum Research Laboratories, The University of Kansas, 2095 Constant Avenue, Lawrence, KS 66047, USA

Carlsson, Enar
Nordreviksvägen 23A, 42737 Billdal, Sweden

Cautreels, W.
Solvay Pharmaceuticals, Brussels, Belgium

Dingermann, Theodor
Institute for Pharmaceutical Biology, Goethe-University, Frankfurt/Main and Center for Drug Research, Development, and Safety (ZAFES), Biocenter, Max-von-Laue-Strasse 9, D-60438 Frankfurt, Germany

Freire, Ernesto
Dept. of Biology and Biophysics, Johns Hopkins University, 3400 North Charles Street, 114 Mudd Hall, Baltimore, MD 21218, USA

Goodford, Peter
The Laboratory of Molecular Biophysics, Rex Richards Building, South Parks Road, Oxford OX1 3QU, England

Haas, M.
Biomade Technology Foundation, Nijenborgh 4, 9747 AG Groningen, The Netherlands

Jourdan, N.
Ariana Pharma, BioTop, Institut Pasteur, 28 rue Dr Roux, 75724 Paris CEDEX 15, France

Jullian, N.
Ariana Pharma, BioTop, Institut Pasteur, 28 rue Dr Roux, 75724 Paris CEDEX 15, France

Kruse, C.G.
Solvay Pharmaceuticals, Weesp, The Netherlands

Kuntz, I.D.
Dept. of Pharmaceutical Chemistry, UCSF, San Francisco, CA 94143-2240, USA

Lange, J.H.M.
Solvay Pharmaceuticals Research Laboratories, C.J. van Houtenlaan 36, 1381 CP Weesp, The Netherlands

Oprea, T.
Division of Biocomputing, MSC11 6145, University of New Mexico School of Medicine, 2703 Frontier NE, Albuquerque, NM 87131, USA

Ottenheijm, Harry C.J.
Department of Pharmacology and Toxicology, Cardiovascular Research Institute, University of Maastricht, P.O. Box 616, 6200 MD Maastricht, The Netherlands

Robillard, G.T.
Biomade Technology Foundation, Nijenborgh 4, 9747 AG Groningen, The Netherlands

Scholtmeijer, K.
Department of Biology, University of Utrecht, Padualaan 8, 3584 CH Utrecht, The Netherlands

Smisterova, J.
Biomade Technology Foundation, Nijenborgh 4, 9747 AG Groningen, The Netherlands

Steinborn, C.
Solvay Pharmaceuticals, Hannover, Germany

Timmerman, H.
Professor emeritus Pharmacochemistry, VU University Amsterdam, De Boelelaan 1105, 1081 HV Amsterdam, The Netherlands

Turski, L.
Solvay Pharmaceuticals, Weesp, The Netherlands

Contents

Conference Preface
and
Keynote Lecture

Towards Drugs of the Future
C.G. Kruse and H. Timmerman (Eds.)
IOS Press, 2008

doi:10.3233/978-1-58603-949-3-3

Conference Preface

The venue of the Solvay Pharmaceuticals Research Conference 2007 "Towards Drugs of the Future" (September 26-28, 2007) was special: the Bavarian mountains near Garmisch-Partenkirchen in Southern Germany. A splendid environment calling for a top performance, culminating in the ascent of the famous Zug Spitze with all conference attendees.

Building upon a tradition, this was already the 9th in the series of Solvay Pharmaceuticals Research conferences, where industry meets academic excellence in the search of novel therapies.

This volume forms the Proceedings of a highly interesting scientific meeting, addressing the multidisciplinary field of Medicinal Chemistry. In the programme a number of critical issues in the fields of design, synthesis and characterization were discussed.

In the theme of *Drug-Receptor Interactions* emphasis was put on targets (Goodford), thermodynamic aspects (Freire) and on structure-based design (Kuntz).

In the session on *Compounds and Synthesis* the subjects of chemical diversity (Oprea) and the interface chemistry-biology (Ottenheijm) got full attention.

The special challenges of designing prodrugs of peptides were introduced by Borchhardt, whereas Afshar's presentation dealt with the complex issue of so-called multiparameter approaches in drug design.

In a session *Pharmacokinetics-ADME* Dingermann discussed the problems caused by terrific compounds which are as medicinal compounds handicapped. Robillard presented in his contribution the possibilities that nanotechnology may offer for making proteins applicable as therapeutics.

The programme was completed by two case stories in drug development. The omeprazole story was told by Carlsson. An intriguing paper on a conceptual design process of cannabis CB_1 receptor antagonists (Lange) concluded the programme.

During the symposium a session was devoted to an interactive discussion on a number of provoking theses; which had been given to "introducers" beforehand. The theses were:

- In 20 years from now drug design and drug screening will be carried out *in silico* completely.
- In 20 years from now all novel medicines will be biologicals.
- In 10 years from now medicinal chemistry will no longer take place within pharmaceutical companies.

The organizers were very pleased by the outcome of the discussions and concluded that such sessions can be very helpful for intensifying interactions between participants of scientific meetings.

This event brought together scientific experts in different disciplines from academia and from the three Solvay Pharmaceuticals research centres in Daix (France), Hannover (Germany) and Weesp (The Netherlands).

We thankfully acknowledge our distinguished speakers as well as all participants for their contributions and also Mrs. Ada Gieling and Mrs. Marjolein Mulder for the excellent organisation of this great scientific and social event.

Henk Timmerman
Chris Kruse

Towards Drugs of the Future
C.G. Kruse and H. Timmerman (Eds.)
IOS Press, 2008

doi:10.3233/978-1-58603-949-3-5

Targets for Therapy Are we Doing Anything Well?

Peter Goodford
The Laboratory of Molecular Biophysics, Rex Richards Building, South Parks Road, Oxford OX1 3QU, England

Abstract. The definition of a 'Target' for drug design; the selection by a pharmaceutical company of a particular Target; and the responsibilities of the pharmaceutical industry to Society as a whole are briefly discussed.

Keywords. Pharmaceutical industry, medicines, cost, safety, responsibility

Introduction

The selection of a Target is probably the most important decision which is ever made in a pharmacological project. However it is a good idea to decide and agree the precise objective of one's research, before even starting to think about any specific Target for therapy. In the pharmaceutical industry this usually means the Company's objective, and different people in a company may have distinct views about this. For instance the investors probably need to optimise their financial returns, but may not be agreed about the time scale on which they are operating. A long-term goal of the scientists may be to develop and extend their own field of research, and of course they also want to find new kinds of therapeutic agent. The clinicians' main aim is always to help sick people, but clinicians give a lot of thought to drug safety, and they always try very hard to protect patients and volunteers from any possible harm. The trade union representatives are also anxious to protect their members from harm, and naturally need to maximise the workers' pay. Everybody's objective is to have a successful Company, but they may not be able to agree about the meaning of "success" or the best way of achieving it. Moreover there is always the influence of "chance", and this can sometimes be more important than almost anything else.

As well as the overall research objective, there are many restraints which must be taken into account before thinking about Targets. The acceptable techniques and procedures for a novel project may be controlled at several different levels of the overall command structure, from the United Nations General Assembly down to a section manager in a sub-department. Moreover the constraints coming from different command levels may be incompatible with each other. For instance the study of human foetal material may be quite acceptable at one level, but absolutely forbidden at another level, or at another time, or in another place.

It is also essential to consider whether religious beliefs, for example, might constrain the choice of objective. A belief may be very strongly held by a group of sincere people who have never seriously considered whether a particular medical research project would genuinely help them, or help their families, or help their neighbours. Other kinds of local pressure groups can demand that research should be done in a particular way, although such groups may lack any official authority, and may have no explicit justification for their sincerely-held views.

It is not always easy to accommodate all these wide-ranging requirements, and one may sometimes be forced to abandon a research project altogether, or postpone it, or move it to a distant location. Clearly the choice of a Target is not a simple decision based solely on clinical and scientific concepts.

1. Cost and Safety

All these needs and pressures add to the overall cost of a medicine. Part of the expense may seem to be carried by the company which initiates the project, but in the final analysis each extra requirement increases the overall price of therapy, and everyone worldwide has to pay something towards that raised cost. Even if an international relief organisation decides to subsidise one particular medicine in one particular place, the net result may simply be that a different aid project goes short. Better and safer medicines always have to be paid for.

Whenever a country considers if it should approve the use of a novel therapeutic agent, their national Authorities may well ask for more tests, or an altered formulation, or different packaging or labelling, or perhaps a modified shelf life. They may claim this is done in the "best interests" of the patient, but every additional demand adds to the overall expense. Furthermore it may sometimes seem as if an extra test (and the consequent extra cost) is actually demanded in the "best interests" of the Authorities rather than the patients, because the Authorities wish to protect themselves from any conceivable accusation of negligence. Companies do not normally alter the properties or presentation of a medicine without good reason, and anything which pushes up the cost is bad in principle because it reduces the number of patients who may benefit.

In fact the situation with therapeutic agents today is somewhat like the situation with cars at the start of the 20th century. Everybody wants therapy and some people can afford to have what they want, but others in the developing world must just watch and wait. Medicines are potentially dangerous (like cars), and Governments therefore feel that they should protect their poorly-informed citizens from the risks of unlimited access to novel therapy. A hundred years ago the British Government had similar worries about motoring, and decided that a man with a red flag should always walk in front of a car to warn pedestrians of the danger. However, there were still road accidents in spite of this well-meant precaution, and so an unintended sequence of rising costs was unconsciously initiated.

The safety of cars (and medicines) has improved since that time, but their cost has also gone up so that many people in the developing world today still need better transport (and better access to medicines) in order to live a reasonably safe and civilised life. Moreover this cycle seems to have no end, because it will always be possible to improve safety as long as people are forced to pay for it, and as long as they can find enough money to pay. We now seem to have a situation in which the safety standards of developed countries drive up the cost of therapy worldwide, and so the developing world cannot afford novel treatments dedicated to the treatment of its particular diseases. People who live in the Third World have to go without new medicines devised for their specific needs, because they cannot pay for the safety standards demanded by developed countries (Figure 1).

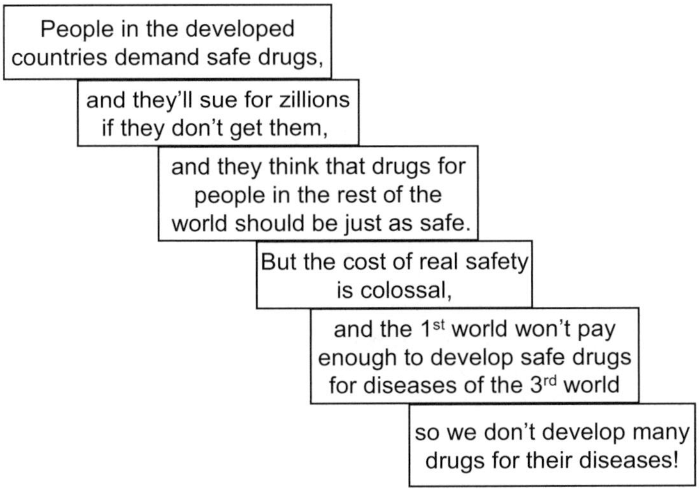

Figure 1. Is this why we don't develop enough drugs for the developing world?

It is said that some AIDS sufferers in California complained very loudly when the first AIDS prophylactics were being assessed. Those patients wanted the American approval process to be speeded up, because they thought they would die slowly and miserably if their therapy was delayed. Perhaps they were an unusually vocal, articulate and well-educated group who knew how to protest effectively, but when they demanded unapproved drugs without delay they were behaving like people in the developing world! They were intelligent patients who loudly and understandably demanded novel therapy as quickly as possible, even if faster delivery meant that some theoretically avoidable dangers might still remain. They just wanted a chance to live!

In fact there are relatively few pharmaceutical companies today which have one of their main laboratories exclusively dedicated to the study of Third World diseases. For example nobody knows how long AIDS has existed in Africa, but it has only been studied seriously since people in developed countries began to suffer and demand effective therapy. We do not target Third World diseases often enough unless we also suffer from them, and this demonstrates rather clearly how the choice of a research Target can be influenced by Costs which are determined to a large extent by arbitrary safety criteria.

The importance of Cost is exemplified in a slightly different situation by the recent introduction of Lucentis therapy for the treatment of Age-Related Macular Degeneration. Some national and local authorities in the United Kingdom do not promote the use of Lucentis because they claim it is not "cost-effective", but it has never been at all easy to establish a reliable and trusted numerical scale for comparing the cost-effectiveness of different medicines. For example, a year's Herceptin therapy is priced at £24,420 by one expert and more than twice as much by another [1,2], and so we should not be surprised if patients dispute "cost-effectiveness" claims when they are rapidly going blind, while people living in a different administrative region already benefit from Lucentis therapy. The Cost of therapy does matter, but how should it be measured reliably?

2. Responsibility

The well-meant and generous decision of governments in developed countries to supply free anti-AIDS drugs to patients in Africa may become a gesture which the whole world will come to regret. This wide-spread distribution of an anti-infective agent, without correspondingly generous provision of trained medical personnel to ensure proper compliance, seems irresponsible because it is almost guaranteed to lead to drug resistance sooner or later. Supplying the drugs seems to be a relatively cheap short-term option for Western governments, compared with the cost of maintaining enough trained staff all over the African continent for the foreseeable future. Everyone must be concerned because people worldwide may eventually suffer in a drug-resistant AIDS epidemic, unless there are enough well-trained people to ensure adequate compliance now. Perhaps it is short-sighted policy for a pharmaceutical company to select a Target for study, if there is a significant possibility that governments may interfere unwisely with the use of the medicines which arise from its use.

In fact the pharmaceutical industry has an outstandingly responsible attitude to the needs and demands of the Society in which it is based, even when those demands are completely unrealistic. People in developed countries expect medicines to be totally safe, although they know that absolute safety is an impossible standard which can never be achieved. The general attitude of Society towards some other industries is entirely different, and it is widely accepted that hundreds of people will be killed or wounded daily in car accidents, whilst a single death due to a therapeutic agent may often be big news.

I believe that the pharmaceutical industry has a wider duty to Society than the passive acceptance of unreasonably over-ambitious demands for perfect safety. In order to deal effectively with this situation it needs to initiate a long-term, well-funded wide-ranging independent programme of psychological research in order to understand why it is judged so much more harshly than other rich and profitable businesses like car manufacture. Why do people get so excited whenever a big and beautiful new car is introduced, but take for granted a wonderful new antibiotic? Why do they expect that the doctor will always be able to cure them with absolutely safe and completely reliable medicines? Why is a new surgical procedure such as liposuction so exciting to the man in the street, while a new therapeutic agent is just boring? It is essential to have the answers to such questions, before we can hope that people will accept a certain low level of risk from therapeutic agents, in the way they already accept a much greater risk from cars.

There is another unexpected, important and perhaps insuperable problem which the pharmaceutical industry urgently needs to address. In spite of funding problems it has successfully ameliorated the health of very many people in the developing world where malaria, for example, has been widely reduced or almost eliminated. Our therapeutic successes led at first to an improving standard of living worldwide, but the population of many countries is now rising so fast that they can no longer feed their people. Rich well-developed countries may still be able to buy the food they need, but the consequences of an ever-growing world population will inevitably become more serious for everybody.

It is the success of the pharmaceutical industry which has led to this situation, and the industry should now receive strong encouragement to search for effective, imaginative, popular and cheap methods of birth control. For example I would like to see compounds which combine a powerful male contraceptive with the properties of Viagra, so that men everywhere would urgently want to take "The Pill" as long as it's price was acceptable. I believe that every baby should now be a wanted baby, and that the pharmaceutical industry should be encouraged to embark on life-style therapy so that people everywhere can enjoy a healthier, happier, sustainable, well-fed long-term future.

3. Targets for Therapy

The Black Death swept across the world in the 14th century. It was "The Will of God", and killed about half the population of Europe. Everyone fled to the countryside if they could afford to flee, including doctors who knew they could do nothing useful because they had no specific Target for their therapy. They just knew that people got ill and died.

If the doctors had known that the Plague was carried by black rats they would have had a Target, and would have been able to recruit and train rat-catchers. If they had known that the rats carried fleas they would have had another Target to attack. If they had known about bacteria they would have had three, but in the absence of any specific Target they could do nothing.

Several hundred years later Lady Mary Wortley Montague was the wife of the British ambassador in Turkey. She was an exceptionally intelligent and beautiful lady (Figure 2), and when there was a serious smallpox epidemic in Constantinople in 1717, she decided to ask the ladies of the harem why they were not as frightened as she was herself. They explained that they had all been "inoculated", and she immediately followed their example and had her children protected in the same way. When she got back to England she told the King, and he thought it was a splendid idea for his Godchildren, but not for his own son who would become King one day! However the British doctors still had no Target to think about, and were naturally opposed to Lady Mary's ideas because they did not understand how inoculation worked.

Figure 2. The Lady Mary Wortley Montague.

Nevertheless medical and scientific concepts were becoming clearer by this time. Of course "The Elements" were still Air, Earth, Fire and Water, but the stream of scholars from Constantinople had turned into a flood after the Ottoman invasion of 1453, and people in Europe had learned from them about the classical concept of atoms. Lagrange's Chemistry first referred to "molecules" in 1800, but nobody had any ideas about macromolecules. William Withering published "An Account of the Foxglove" in 1785, and this was the first modern book on pharmacology. He practised medicine not so very far from Oxford, and noticed that some village folk used foxglove leaves in order to treat "dropsy" which was their name for congestive heart failure. Soon afterwards "vaccination" was introduced by Edward Jenner who showed that cowpox gave good protection against smallpox without the risks which were associated with Lady Mary's inoculations. These were all critically important advances, but although Withering and Jenner were exceptionally good scientists, they knew nothing about Targets and did not fully understand why or how their therapy worked.

Everything changed in the nineteenth century with the concept of an infectious organism having a structure that one could think about, and the complementary concept of a molecule which could be used to attack that organism. Clinicians and scientists now had a Target, and they had a Bullet which they could shoot at it. Very soon they would have multiple Targets and many Bullets, although today's apparently wide and endless river of safe, potent and effective therapeutic agents did not become a reality until sulphonamides were generally introduced in the 1940s.

In the 21st century we can look back at all this slow and rather haphazard progress, and may be contented with the feeling that we have reached the end of "The Road". We now have crystal structures for our protein, glycoprotein, DNA and other macromolecular Targets, and are tempted to believe that we only need to choose one of these for detailed study in order to have unlimited successes. I believe that nothing could be further from the truth, and that our ideas about Targets will continue to evolve for the foreseeable future. "The Road" was already dividing in the 1970s or thereabouts, and we now have several paths to follow.

In the 1960s, when I decided to use Max Perutz's observed haemoglobin structure in order to design small drug molecules, less than half a dozen macromolecular X-ray structures had been observed anywhere, and most of them were unsuitable as Targets for therapy. It had taken Perutz half a lifetime to determine his structure which had four subunits, but he most generously gave us his coordinates and his Target [3] was used to design many novel haemoglobin ligands including a potent anti-sickling compound [4]. By that time a smaller protein (Dihydrofolate Reductase) had become a popular enzyme to study for drug design, but the trend towards ever smaller Targets then began to go into reverse. Targets had shrunk from rats to fleas to bacteria to big and finally to small proteins, but today we are starting to look again at bigger Target systems because we can observe the structure of viruses and ion channels and cell membranes. I believe that today's young scientists may eventually be designing drugs by solving the Schrödinger equation for a bacterium, or perhaps for a tapeworm. Macromolecular structures observed by X-ray crystallography are not the last word in Targets, but just a recent stage in our attempts to find new therapeutic agents.

Crystal geometry seriously restricts molecular motion, and only permits us to view a static model with hints of mobility. New methods have therefore come forward which allow us to study the mobility and ionisation of a macromolecule, and observe many related properties of the surrounding solvent and solutes. We are moving from a world in which NMR practitioners competed with crystallographers, towards a new environment in which these techniques are used collaboratively with each other, and in combination with microcalorimetry, mass spectroscopy, molecular mechanics, electron spin resonance, computational and many other relatively novel methods.

It is too soon to predict where these new ways of studying Targets may lead us, and traditional X-ray structures will still be used for a long time in order to design drugs. For instance there is a cleft in the haemoglobin macromolecule which could be used to find compounds for the treatment of anaemia, heart failure and circulatory diseases, and this Target has many other important advantages. Human haemoglobin and human blood could be studied giving results directly relevant to man so that fewer animal experiments would be required, and small anionic haemoglobin ligands might diffuse freely into erythrocytes, but be excluded by the transmembrane potential from other cells where they could have unwanted side effects or cause toxicity.

It is important to bear such aspects and many more questions in mind when deciding on a Target. What regulatory controls and ethical matters must be considered? How many patients will benefit? How relevant and trustworthy and cost-effective are our preclinical and clinical models? How safe is safe enough? What is the time scale? Who makes the final choice of Target? How is society to pay for its new medicines?

4. Conclusions

It is not easy to define objectively how safe a novel therapeutic agent should be, and perhaps that is why I have not been able to write this short review as objectively as I would wish. Each pharmaceutical company clearly needs to have its own definition of "Target" and a preferred method for choosing it. For instance it could choose:

(1) By objective: "We are going to find the cure for cancer!"
(2) By copying: "We will look at thrombin (like everyone else)."
(3) By development: "We will find a safer non-sedating antihistamine than…."
(4) By facilities: "Now that we have a nearby synchrotron X-ray source….."
(5) By pathway: "We will study folate metabolism."
(6) By mechanism: "We will exploit our knowledge of Cell Cycle Control."
(7) By ligand: "We will find and study this new hormone…."
(8) By structure: "We will study the influenza neuraminidases…."

Some of these methods are more likely than others to yield worthwhile novel therapeutic agents, and each company must make the decision for itself. I believe its choice may determine its future success, and perhaps its continued long-term existence.

References

[1] K. Sikora. Paying for cancer care – a new dilemma. *J. R. Soc. Med.* **100** (2007) 166-169. Comment in: *J. R. Soc. Med.* **100** (2007) 303.
[2] M. Atkins. Cost to the NHS of Roche oncology treatments. *J. R. Soc. Med.* **100** (2007) 303.
[3] M.F. Perutz. The haemoglobin molecule. *Proc. Roy. Soc. B.* **173** (1969) 113-140.
[4] G. Kneen and R.D. White. BW12C: a new anti-sickling agent. *Br. J. Pharmac.* **74** (1981) 965P.

Towards Drugs of the Future
C.G. Kruse and H. Timmerman (Eds.)
IOS Press, 2008

doi:10.3233/978-1-58603-949-3-13

A Thermodynamic Platform for Drug Discovery and Optimization

Ernesto Freire
Dept. of Biology and Biophysics, Johns Hopkins University, 3400 North Charles Street, 114 Mudd Hall, Baltimore, MD 21218, USA

Abstract. The application of newly developed thermodynamic guidelines for drug development has yielded unprecedented results in the design and optimization of drug candidates against important diseases. By experimentally monitoring not only binding affinity but the thermodynamic forces that determine binding, enthalpy and entropy, it is possible to significantly accelerate the development process, as these forces reflect different types of drug-target interactions. A thermodynamic platform allows faster potency and selectivity optimization in conformity with existing rules for the design of drug-like molecules and oral bioavailability.

Keywords. Binding affinity, isothermal titration calorimetry, ITC, lead optimization, thermodynamics

Introduction

During the past decades, drug discovery has heavily relied in the screening of large libraries of compounds. Usually, compounds identified using this approach ("hits") have binding affinities in the low to high micromolar range, requiring potency improvements of five or more orders of magnitude before they can become viable drug candidates.

Once hits are identified, two important questions need to be answered: which ones should be selected for optimization and, once the selection is made, how the optimization is performed. The selection process involves different areas, ranging from potency, selectivity, bioavailability and toxicity, even including intellectual property issues. In the area of potency and selectivity, the subject of this presentation, important questions include: Which compounds have the best potential to be optimized? What is the best strategy for potency and selectivity optimization? What type of chemical functionalities are needed and where? The correct answers to those questions are of the outmost importance since the wrong decision might lead to a dead end rather than to a successful drug candidate.

Forces that Determine Potency and Selectivity in Drug Binding

Different forces determine binding potency. These forces can be classified as attractive forces between the drug molecule and the target and repulsive forces that drive the drug away from the solvent. Among the attractive forces, Van der Waals and hydrogen bonds are the most prominent. The strength of Van der Waals interactions depends on how closely the ligand molecule fits within the binding pocket (shape complementarity), while the strength of hydrogen bonds depends on the distance and angle between hydrogen-bond acceptors and donors in the drug and target molecules. Among the repulsive forces between the drug and solvent, the hydrophobic effect is the most important and essentially depends on the burial of hydrophobic groups from the solvent, with no significant distance or angular constraints. On the contrary, the hydrogen-bonding constraints are so strict that many times weak hydrogen bonds end up opposing rather than favoring binding, as desolvation penalties and compensatory entropic effects predominate. A hydrogen bond made against a relatively unstructured group in the protein (e.g. a solvent exposed side chain) will carry a conformational entropy penalty. For these reasons, designing good hydrogen bonds for potency optimization has been a difficult task.

In addition to potency, a major roadblock in optimization is selectivity. For example, kinases are major targets in drug discovery. All of them bind ATP and have homologous binding sites. How do you hit a particular one or class and not the others? While all the forces mentioned above contribute to potency, not all of them contribute equally to selectivity. A tight fit between the drug molecule and binding pocket maximizes Van der Waals interactions and lowers the probability that the drug will accommodate equally well in other protein. Hydrogen bonds are major contributors to selectivity due to their stringent distance and angle constraints. A different arrangement of donors and acceptors, as expected to be the case in proteins other than the target, not only weakens or eliminates the favorable energy of hydrogen bonds, but leaves intact the severe desolvation penalty that originates from the burial of unsatisfied polar groups, thus lowering the affinity towards unwanted targets. Hydrophobicity, not being an attractive force to the target and being only dependent on the burial of non-polar groups from the solvent, does not contribute to selectivity.

There are basically four binding forces that can be optimized during lead optimization in order to improve the affinity and selectivity of the drug candidate: 1) better Van der Waals interactions by improving the geometrical fit of the drug within the binding pocket; 2) the formation of strong hydrogen bonds; 3) increased hydrophobicity; and, 4) minimization of conformational entropy losses by conformationally constraining and pre-shaping the drug molecule to the target.

Binding Thermodynamics

The binding affinity is determined by the Gibbs energy of binding (ΔG), which itself is a function of the enthalpy (ΔH) and entropy (ΔS) changes:

$$\Delta G = \Delta H - T\Delta S$$

Different forces contribute differently to ΔG, not only in magnitude but also in their enthalpic/entropic character [1,2]. Consequently, knowing the enthalpy and entropy changes as well as the changes in heat capacity, ΔC_p, (determined from the temperature dependence of the enthalpy change) provides a unique window into the binding process. The Gibbs energy includes not only terms directly related to the drug-target interaction, which are the ones critical during drug optimization, but also indirect effects like conformational changes of the protein upon binding or coupled protonation reactions. In the vast majority of situations those terms constitute a constant background since the drug molecule under optimization targets the same site and elicit the same inhibitory effect. Exceptions involve the design of inhibitors aimed at modulating an allosteric effect in which conformational changes need to be monitored during optimization [3].

The binding forces that can be manipulated during optimization contribute differently to the enthalpy and entropy changes. Van der Waals interactions and hydrogen bonds contribute favorably to the binding enthalpy, while hydrophobicity contributes primarily to a favorable binding entropy. The introduction of conformational constraints aiming at pre-shaping the ligand to the target minimizes unfavorable conformational entropy effects. Changes in these forces are reflected in the thermodynamic signature of the compound as it moves along the optimization path. A negative heat capacity change is associated with the burial from the solvent of non-polar groups [4].

The thermodynamic signature provides an instantaneous visual representation of the enthalpic and entropic contributions to the binding energy or the differences in binding thermodynamics between one molecule and another. It allows immediate visualization of the relative magnitude of the binding forces. Usually, the original compound or initial hit is taken as the reference and newer compounds are analyzed by plotting the difference in thermodynamic parameters with respect to the reference ($\Delta\Delta G$, $\Delta\Delta H$ and $-T\Delta\Delta S$). An example is shown in Figure 1. In this figure, the thermodynamic signature of darunavir, one of the newly approved HIV-1 protease inhibitors is compared to that of indinavir, one of the first-generation inhibitors (data from this laboratory [5]). Darunavir exhibits much higher potency and improved drug resistance profile. As seen in the figure, the improved potency (75-fold or 2.6 kcal/mol) originates from a much larger favorable binding enthalpy. Indinavir, like other first-generation inhibitors [6], is characterized by unfavorable binding enthalpies consistent with the presence of weak or unsatisfied hydrogen-bonding groups in which the large desolvation enthalpy is the dominating term. These inhibitors bind to the protease in a process dominated by extremely large favorable binding entropy, consistent with a hydrophobically driven binding process (Figure 1). Darunavir, on the other hand, is characterized by extremely favorable enthalpy consistent with an efficient optimization of the drug-target interactions and a minimization of desolvation penalties. As is usually the case in strongly enthalpic compounds, the binding entropy of darunavir is significantly weaker than that of indinavir, a fact that can be attributed to the additional structuring caused by strong hydrogen bonds and a smaller hydrophobic surface burial. It must be noted that darunavir has 3 hydrogen-bond donors, 10 hydrogen-bond acceptors, a ClogP of 2.23 and molecular weight of 547.7. Indinavir has 4 hydrogen-bond donors, 9 hydrogen-bond acceptors, a ClogP of 2.39 and molecular weight of 613.8 (source: chemical database NIAID). These values indicate that dramatically different thermodynamic signatures can be achieved with compounds that have rather similar chemical descriptors.

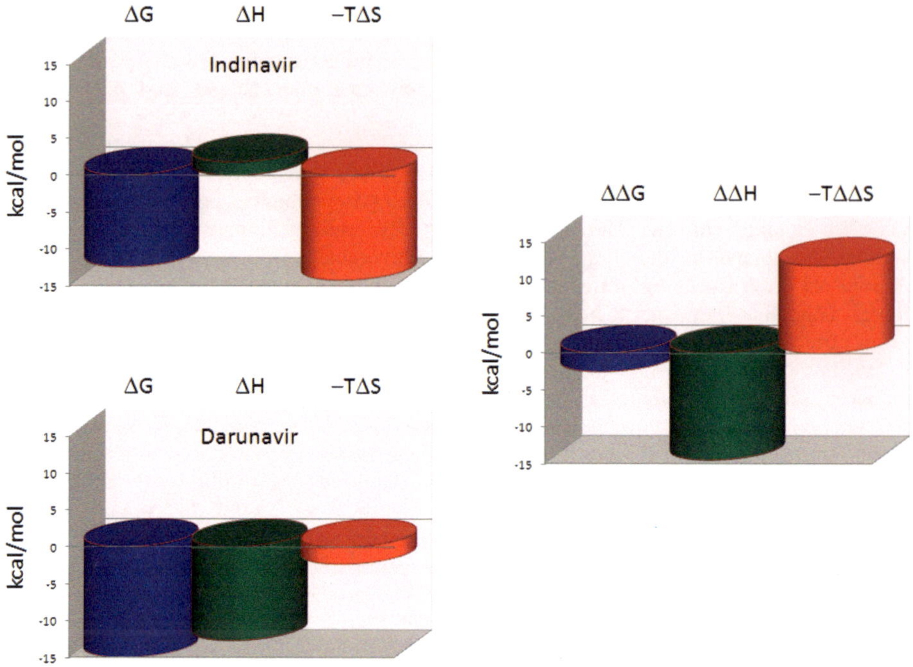

Figure 1. The thermodynamic signatures of the HIV-1 protease inhibitors indinavir (first generation) and darunavir (current generation) are shown on the left side. On the right side, the thermodynamic signature difference is shown. The difference provides a rapid assessment of the improvement in binding energy (ΔΔG) and the origin of that improvement (ΔΔH or -TΔΔS). During lead optimization, a similar plot permits determination of the required actions for further optimization, i.e. better enthalpic or entropic interactions and the necessary chemistry to bring about those changes. In lead optimization, the changes in the thermodynamic parameters are expressed in relation to those of the initial compound in the series, which serves as reference.

Thermodynamic Signatures and Optimization

Thermodynamic signatures measured by isothermal titration calorimetry (ITC) can be used to assess the relative contributions of different forces to the binding affinity of hits identified in enzyme inhibition screens. Compounds with a larger hydrophobic component will exhibit larger binding entropies while compounds with better Van der Waals or hydrogen bonds will exhibit more favorable binding enthalpies. The hit profiling obtained by the thermodynamic signature goes beyond a simple potency ranking since it also provides important insights into the binding mechanisms of different compounds. This information is critical since it could be preferable to select for further optimization a compound with a more robust binding mechanism rather than the hit with the highest affinity. After all, the binding affinities of even the stronger hits are usually very weak.

Once a molecule or scaffold is selected for optimization, its thermodynamic signature provides an accurate blueprint for optimization. The thermodynamic signature permits to expand the SAR process to two additional dimensions (enthalpy and entropy) in addition to potency [7,8]. This information is critical for fast optimization since it indicates the type of changes, enthalpic or entropic, and the necessary chemistry that is required to improve potency and selectivity.

The binding affinity of a compound can be improved by engineering a more favorable binding enthalpy, more favorable solvation entropy, or by conformationally constraining the compound and minimizing the unfavorable conformational entropy change. Obviously, extremely high affinity is achieved when the three factors are optimized simultaneously. However, historically it has been easier to optimize the entropy than the enthalpy. Since the major favorable contributor to favorable entropy is the hydrophobic effect, which is proportional to the number of non-polar groups that are buried from the solvent, the tendency throughout the years has been towards an increase in the hydrophobicity of drug candidates [9]. Medicinal chemists have long learned to conformationally constrain and pre-shape molecules to the geometry of the binding site, which completes the entropy optimization.

There is a limit, however, to the number of hydrophobic groups that can be added to a compound before it becomes completely insoluble and useless as a drug molecule. At some point in the optimization process, it becomes necessary to introduce favorable enthalpic interactions if the goal is to achieve nanomolar or sub-nanomolar affinities. Compounds that exhibit extremely high affinity have been shown to display both favorable entropic and enthalpic interactions [5,10]. Enthalpic interactions are not introduced by increasing the polarity of a compound but by the correct placement of hydrogen-bond donors and acceptors. Poor or unfavorable binding enthalpies are usually associated to an incorrect location of those groups rather than fewer of those groups.

The optimization of the binding enthalpy has been notoriously more difficult than the optimization of the binding entropy. Polar groups carry a large unfavorable enthalpy of desolvation, close to one order of magnitude larger than that of non-polar groups [11] (Figure 2). A polar group needs to establish a very good interaction with the target in order to compensate for the desolvation penalty and make a favorable contribution. For this reason, they are often engineered as solubilizers of otherwise extremely hydrophobic compounds rather than major contributors to affinity.

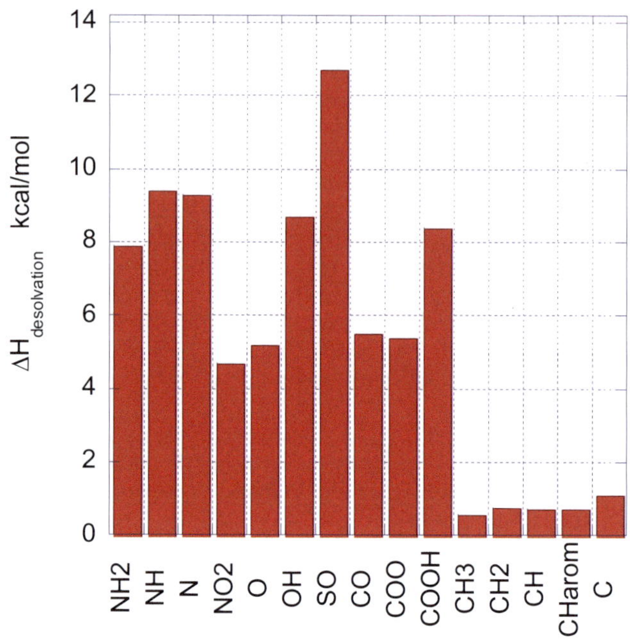

Figure 2. The desolvation enthalpy of different chemical functionalities. Data has been taken from Cabani *et al* [11] and represent values from water to the gas phase.

Introducing Hydrogen Bonds for Potency and Selectivity

During optimization it is common to observe that the potency of a compound does not improve after the introduction of a hydrogen-bond donor or acceptor moiety. It could be that the hydrogen bond is not made or that the enthalpy associated with the formation of the hydrogen bond is compensated by opposing entropy, resulting in no potency gain. The thermodynamic signature provides an instantaneous account of the situation.

Figure 3 provides an illustration of a situation encountered in this laboratory during the optimization of plasmepsin inhibitors. This figure illustrates the enthalpic consequences of introducing a hydroxyl group in the benzyl moiety of the compound. As seen in the figure, addition of the hydroxyl group at the meta-position increases the binding enthalpy by -2.8 kcal/mol clearly indicating a strong hydrogen-bond interaction. Addition of the hydroxyl group to the para-position further increases the favorable enthalpy by an additional -0.4 kcal/mol. As a control, it can be seen that the addition of a methyl group at the same para-position not only does not improve the favorable enthalpy but reduces it by 1.2 kcal/mol over that of the parent compound. This example clearly illustrates the ability of ITC to identify and find the best location for a hydrogen-bonding moiety. The identification of the best location to introduce hydrogen-bond donors or acceptors can be readily accomplished with the guidelines provided by the thermodynamic signature.

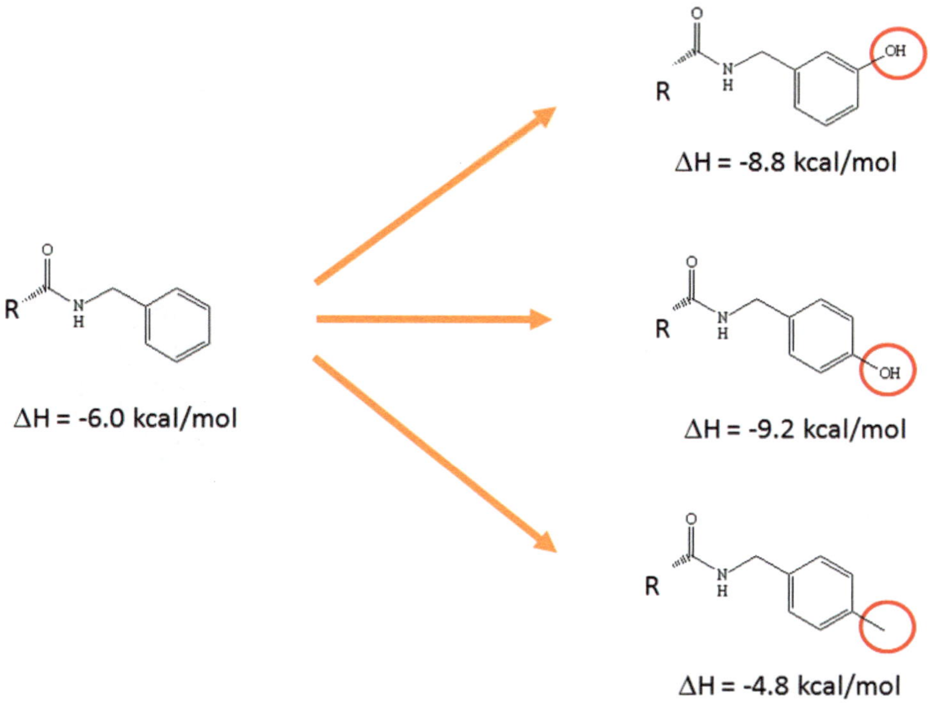

Figure 3. The binding enthalpy opens a new dimension to SAR. In this particular example, the enthalpy serves to identify that the hydroxyl group is participating in a hydrogen-bonding interaction, and furthermore, to identify the best location.

A different situation often encountered in lead optimization is that of a hydrogen bond that is formed but does not contribute to the binding potency. Figure 4 illustrates this situation with a pair of HIV-1 protease inhibitors (KNI-10033 and KNI-10075) that differ at only one position [12]. KNI-10033 is a potent experimental HIV-1 protease inhibitor with picomolar affinity against the wild-type enzyme (Kd = 13pM). The potency of the inhibitor is the result of favorable enthalpic (ΔH = -8.2 kcal/mol) and entropic ($-T\Delta S$ = -6.7 kcal/mol) interactions. The replacement of the thioether group in KNI-10033 by a sulfonyl group (KNI-10075) results in a strong hydrogen bond with the amide of Asp 30B of the HIV-1 protease. This additional hydrogen bond improves the binding enthalpy by -3.9 kcal/mol; however, the enthalpy gain is completely compensated by an entropy loss, resulting in no affinity change. Crystallographic and thermodynamic analysis of the inhibitor/protease complexes indicated that the entropy loss originates from a conformational entropy loss due to the structuring induced by the hydrogen bond and a smaller favorable desolvation entropy [12].

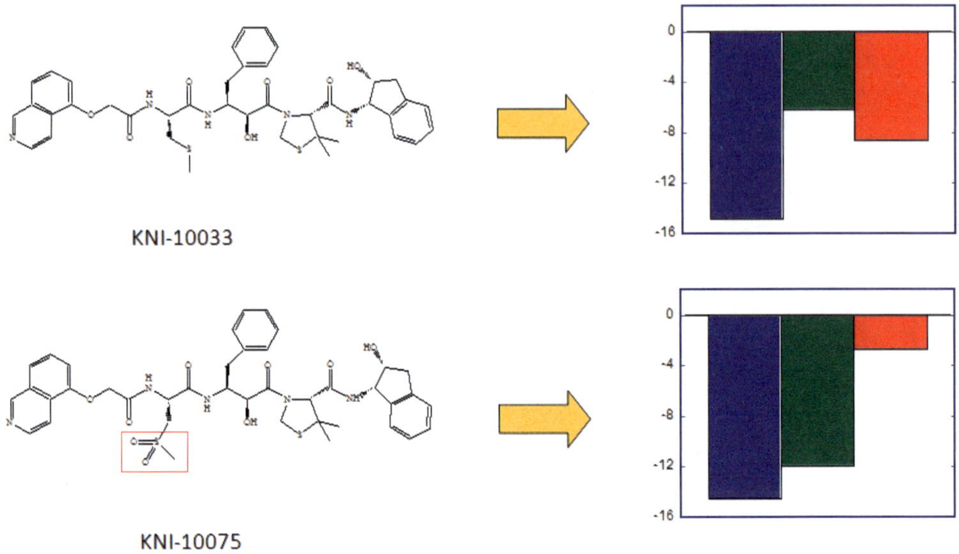

Figure 4. Thermodynamic signatures of two HIV-1 protease inhibitors that differ at only one position. The additional hydrogen bond in KNI-10075 increases significantly the favorable enthalpy of binding but has no effect in binding potency because the enthalpic gain is compensated by an entropic loss.

The engineering of hydrogen bonds for improved binding affinity requires considerations other than donor/acceptor distances or angles. It is apparent that the structuring associated with hydrogen-bond formation can significantly compensate expected improvements in binding affinity. Consequently, optimization of the contribution of hydrogen bonds to binding affinity requires minimization of any structuring effect. The minimization of the structuring effect can be achieved by directing the hydrogen bond towards well-structured regions of the protein and/or by placing the acceptor/donor functionalities in regions of the compound that are already conformationally constrained in the unbound state. On the contrary, targeting hydrogen bonds to unstructured regions will carry the maximal structuring penalty.

Another detrimental effect is a lower degree of desolvation originating from the structural constraints imposed by the hydrogen bond that prevents full accommodation of neighboring groups to maximize burial from solvent. According to the results of Lafont *et al* [12], for the case of KNI-10075, desolvation accounted for 35% of the total entropy compensation effect, while the majority (65%) was accounted for by conformational entropy losses.

Isothermal Titration Calorimetry

Isothermal titration calorimetry (ITC) is the only technique that is able to determine the Gibbs energy, association constant, enthalpy and entropy changes, as well as the stoichiometry of binding. Furthermore, ITC does that in just one experiment, without the need for labels, and without the need to immobilize either the protein or ligand [13-15]. Experiments at different temperatures allow determination of the heat capacity change, and experiments utilizing buffers with different ionization enthalpies allow determination of any protonation/deprotonation coupled to the binding reaction [16].

ITC measures the heat released or absorbed by the stepwise addition of a ligand molecule into the reaction cell of the calorimeter containing a solution of the target protein, while maintaining constant temperature and pressure. The content of the sample cell is continuously stirred to ensure proper mixing of the reactants. The heat released or absorbed after each injection is obtained by calculating the area under each peak. Since the measured heat is proportional to the increase in bound macromolecule concentration, its magnitude decreases as the protein becomes saturated. After saturation, additional injections produce small identical peaks, due to non-specific phenomena such as dilution and mixing effects, whose heat effect must be subtracted from all the injection peaks before performing data analysis. Data analysis of the heat effect for each injection leads to the determination of the thermodynamic binding parameters. A typical ITC experiment is shown in Figure 5.

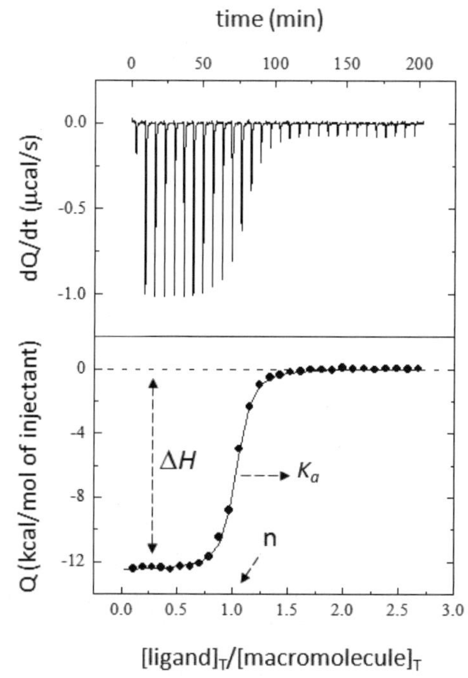

Figure 5. Typical ITC titration of a protein by a small ligand. The signal recorded by the instrument is shown in the top panel. This signal is equal to thermal power required to maintain a constant temperature in the sample cell during the sequence of injections of the ligand into the protein. Depending on the exothermic or endothermic character of the binding reaction, the peaks will show a downward or upward deflection respectively. The area under each peak (integrated heat) for the sequence of injections is shown in the bottom panel. The thermodynamic binding parameters (binding affinity, binding enthalpy and stoichiometry) are obtained through non-linear regression of the experimental data.

Conclusions

The binding potency is determined by enthalpic and entropic forces. In the past, structure-activity relationships (SAR) aimed at improving the affinity of potential drug candidates have been hindered by the absence of information regarding the origin of advantageous or detrimental effects resulting from the introduction of certain chemical functionalities. The thermodynamic signature determined by ITC introduces two additional dimensions to SAR, thus providing clear guidelines for a faster lead optimization.

References

[1] E. Freire. A thermodynamic guide to affinity optimization of drug candidates. In: G. Waksman (Eds): *Proteomics and Protein-Protein Interactions: Biology, Chemistry, Bioinformatics and Drug Design*. Springer NY, USA (2005) 291-307.

[2] A.J. Ruben, Y. Kiso, E. Freire. Overcoming roadblocks in lead optimization: a thermodynamic perspective. *Chem. Biol. Drug Des.* **67** (2006) 2-4.

[3] A. Schön, N. Madani, J.C. Klein *et al.* Thermodynamics of binding of a low-molecular-weight CD4 mimetic to HIV-1 gp120. *Biochemistry* **45** (2006) 10973-10980.

[4] I. Luque and E. Freire. Structural parameterization of the binding enthalpy of small ligands. *Proteins* **49** (2002) 181-190.

[5] H. Ohtaka and E. Freire. Adaptive inhibitors of the HIV-1 protease. *Prog. Biophys. Mol. Biol.* **88** (2005) 193-208.

[6] A. Velazquez-Campoy, Y. Kiso, E. Freire. The binding energetics of first- and second-generation HIV-1 protease inhibitors: implications for drug design. *Arch. Biochem. Biophys.* **390** (2001) 169-175.

[7] E. Freire. Isothermal titration calorimetry: controlling binding forces in lead optimization. *Drug Discov. Today 1* (2005) 295-299.

[8] R.W. Sarver, J. Peevers, W.L. Cody *et al.* Binding thermodynamics of substituted diaminopyrimidine renin inhibitors. *Anal. Biochem.* **360** (2007) 30-40.

[9] C.A. Lipinski. Physicochemical properties and the discovery of orally active drugs: technical and people issues. In: M.G. Hicks and C. Kettner (Eds): *Proceedings of the Beilstein Institute Workshop. Molecular Informatics: Confronting Complexity*. Logos Verlag Berlin, Germany (2003) 59-78.

[10] T. Carbonell and E. Freire. Binding thermodynamics of statins to HMG-CoA reductase. *Biochemistry* **44** (2005) 11741-11748.

[11] S. Cabani, P. Gianni, V. Mollica *et al.* Group contributions to the thermodynamic properties of non-ionic organic solutes in dilute aqueous solution. *J. Solution Chem.* **10** (1981) 563-595.

[12] V. Lafont, A.A. Armstrong, H. Ohtaka *et al.* Compensating enthalpic and entropic changes hinder binding affinity optimization. *Chem. Biol. Drug Des.* **69** (2007) 413-422.

[13] T. Wiseman, S. Williston, J.F. Brandts *et al.* Rapid measurement of binding constants and heats of binding using a new titration calorimeter. *Anal. Biochem.* **179** (1989) 131-135.

[14] A. Velazquez-Campoy and E. Freire. ITC in the post-genomic era...? Priceless. *Biophys. Chem.* **115** (2005) 115-124.

[15] J.M. Brandts, R.K. Brown, R. O'Brien *et al.* ITC-derived binding constants: using microgram quantities of protein. In: M. Cooper (Eds): *Label-Free Biosensors: Techniques and Applications*. Cambridge University Press (2007).

[16] B.M. Baker and K.P. Murphy. Evaluation of linked protonation effects in protein binding reactions using isothermal titration calorimetry. *Biophys. J.* **71** (1996) 2049-2055.

Towards Drugs of the Future
C.G. Kruse and H. Timmerman (Eds.)
IOS Press, 2008

doi:10.3233/978-1-58603-949-3-23

Structure-Based Design: Past, Present, and Future

I.D. Kuntz
Dept. of Pharmaceutical Chemistry, UCSF, San Francisco, CA 94143-2240, USA

Abstract. This overview presents the Structure-Based Drug Design, with a focus on what can be accomplished and its long-term potential. While database mining has repeatedly been able to identify compounds in the Available Chemical Directory (ACD) that inhibit enzymes and allosteric systems in the low micromolar range, it is only with extensive chemical efforts that nanomolar inhibitors have been developed. Merging combinatorial chemistry strategies with structure-based design principles has had a major impact on our ability to identify sub-nanomolar inhibitors with the selectivity and pharmaceutical properties to be plausible preclinical candidates.

Keywords. Structure-based design, rational drug design, docking, optimized ligands

1. Past

In the late 1970s, it occurred to several researchers that the advent of protein structures from X-ray crystallography would open a new window in the development of drugs. These included two of us present in this conference, Peter Goodford and myself. But there were others of note including Max Perutz and Dan Santi. The vision was straightforward: given the structure of a potential drug target, one selected or "designed" an organic molecule that would provide excellent steric and chemical complementarity to one or more features of the target. Peter's work, and rather quickly thereafter, ideas from Peter Kollman, Martin Karplus and others, stressed the importance of physical interactions, calculated from first principles, that represented the differential attractiveness of one ligand compared to another. Our initial approach emphasized the importance of the geometric matching of ligand and target - the three-dimensional jigsaw puzzle aspect of binding.

There followed many years of exciting developments with a flood of new ideas and approaches and considerable sophistication of physical theory and computer programs. These efforts continue today with dozens of docking schemes in the literature and hundreds of projects in both academic and industrial laboratories using these techniques [1].

There were well-recognized flaws in the early ideas that have still only been partially resolved. These include the difficulty in computing entropic contributions to the free energy of interaction, the need for a high level of accuracy (ca. 0.1 kcals/mole!) to compare with experiment, the conformational plasticity of ligand and target, the sometimes critical role of water molecules, and the difference in difficulty of achieving agonists compared to antagonists. It also quickly became apparent that there was a large gap between discovery of novel ligands and generation of effective lead compounds.

Much of the last decade has been spent dealing with these issues, and most remain active areas of current research, as can be seen in this volume.

It is fair to say that structure-based design is now an accepted tool that has taken its place with other technologies and is actively utilized in all major pharmaceutical companies, as well as many academic and biotech laboratories.

2. Present

The initial DOCK protocol [2] can be summarized as follows:

1. obtain one or more structures of the target;
2. use a library of templates or shapes for the ligands;
3. rank order the library in terms of quality of steric complementarity to one or more target sites;
4. in consultation with the medicinal chemists and pharmaceutical scientists, expand the templates by exploring atom substitutions or fragment additions as well as conformational flexibility;
5. "score" the expanded library using the best available force fields and molecular computations;
6. obtain or synthesize the library and test it;
7. obtain new structures of the target with the effective ligands;
8. repeat the process in an iterative fashion.

A modern protocol would be quite similar but there would be large-scale enhancements at each stage. For example, it is now possible to obtain excellent macromolecular structures via NMR and to make use of all structural data using homology-modeling techniques. The latter, especially, have greatly expanded the number of structures available. Crystallographic methods have greatly improved. Proteins, particularly, are now of very high purity and structures are often available at outstanding resolution (<1.2 Å). Further, the high-resolution structures often are analyzed to show multiple occupancies, allowing the visualization of alternative binding modes.

Computational methods such as molecular dynamics, coupled with Dan Koshland's "induced-fit" have moved us away from the idea that a macromolecule has a single rigid structure. Rather, we now think in terms of a thermodynamic ensemble of structures, amongst which a subset interact most strongly with a particular flexible ligand.

A modern library of ligands would include both expansion and filtering techniques. A compound list would be expanded to include tautomers, charged variants, and alternative conformers. It would be filtered to remove compounds with poor pharmaceutical pedigrees or those that offered known synthetic difficulties. It is now known that there are important numbers of promiscuous inhibitors that, interestingly, confound both computationally and high throughput experimental screens. These can also be filtered out to some extent.

Evaluating the strength of interaction of a ligand with a target (often called "scoring") currently can be carried out at several levels, with the typical trade-off being higher accuracy requiring exponentially more computational effort [3]. The best available technology uses free energy perturbation techniques. This approach requires intensive preparation of the target structure (complete protonation state analysis, proper counter-ions, placement of critical water molecules), and many hours of processor time per ligand, but it can achieve better than 0.5 kcals/mol for a binding free energy. These methods should produce free energy values that do not require scaling or additive constants. There are more approximate simulations such as Linear Interpolation or Molecular Mechanics Poisson Boltzmann that can yield similar accuracy for some systems. Often, however, these methods yield good rank orders at about 1/10 the computational effort but have unexplained offsets in absolute terms. Standard molecular dynamics has difficulties producing free energies. It can be used to estimate enthalpies of interaction.

Because of the large computational needs of direct simulation, most projects use empirical scoring methods that have adjustable constants and heuristic factors. While these approaches often identify interesting hits, their reliability is limited to rank ordering within families of similar compounds. And, of course, they still require proper identification of tautomers and charged states.

Another area of current interest is coupling the docking calculations to synthetic strategies so that the virtual libraries under study can be readily translated to new compounds in the laboratory [4]. Such coupling can be done by individual chemists who wish to explore specific modification reactions or at a company-wide level by codifying chemical procedures - especially combichem approaches based on electronic notebooks and other strategies. One approach that has attracted much attention is fragment linking [5]. Here, a library of fragments is explored, using by NMR or crystallography, and then the fragments are linked in a combinatorial fashion aided by computational simulation. Several companies have indicated that these methods can provide rapid development of leads.

Industry has made excellent use of co-crystallization to "close the loop" for compound optimization. A current structure-based design project could easily include more than 100 crystal structures with the new ligands examined on a daily or weekly basis. Such intense monitoring often clarifies the "mysterious" jumps or losses in activity as changes in binding geometry.

In brief, a combination of good targets, appropriate starting compounds, facile chemistry, crystallography, and computation often allows rapid progress from "hits" to ca. nanomolar "leads". The burden then falls on the pharmaceutical development team and the biologists to generate efficacy in patients with acceptable side effects. The ability to simulate the "ADME" properties or differential binding to a number of adventious targets is relatively limited although there has been good progress on the "physical" properties such as solubility [3] and, to some extent, in predicting metabolism. The choice of molecular target also remains a challenging issue. The difficulty is normally not with the fundamental biochemistry/biology model, but rather the larger systemic context: are there alternative pathways when the target is blocked, how rapidly can drug resistance arise, are there inherent side effects associated with that particular target? It has frequently proven difficult to resolve these questions in preclinical trials, leaving them to be answered at great expense during early phase clinical studies.

Current expectations from using computational screening, while strongly dependent on the compound library, would be ca. 1-5% "hit" rate in the low micromolar range against enzyme targets. For more difficult targets, such as protein-protein interactions or nucleic acid systems, one expects a lower "hit" rate and ca. 10-100 micromolar results. Also note that many of the compounds selected in such screens are not classical competitive inhibitors although they might still show dose-dependent behavior. Compound optimization using computational methods can be expected to improve binding 10- to 100-fold. As noted above, it is generally feasible to estimate the solubility and blood levels of a series of related compounds as long as the mechanism of uptake and distribution does not change along the series.

An interesting idea to assist compound optimization strategies uses "rules of thumb" from the maximal affinity of ligands [6]. This work, built on ideas from Peter Andrews and extended recently by Hopkins [7] suggests that each non-hydrogen atom of a ligand can contribute up to 1.5 kcals/mol of binding. Thus addition of single methyl groups might be expected, optimally, to produce a 10x improvement in binding, while a perfectly positioned phenyl ring could provide upwards of 1000x enhancement. Such dramatic changes are seen experimentally in the very best ligands such as biotin. They are less often seen in medicinal chemistry but can provide a guide to evaluating drug development programs.

3. Future

Let us look ahead. Each of the modules of the structure-based design paradigm is likely to undergo significant change.

First, the target structural databases will increase in a major way in size and also in quality. It is easy to foresee that even the difficult systems of membrane-bound proteins and conformationally flexible RNA targets will yield to improved purity, screening techniques, and engineered mutations. With the continued interest in proteomics and high throughput structure projects, it is safe to predict that there will be some structures for every major class of pharmaceutical target in the next 10-20 years. Further, we will see important enhancements in the homology-based approaches [8]. Of special interest will be efforts to understand the correlated motions and dominant conformations of enzymatic and switching systems and the modeling of protein-protein and protein-nucleic acid complexes [9]. Much of this insight can be expected to arise from careful simulations and homology models using molecular dynamics and brownian dynamics techniques. The result will be a much improved representation of individual proteins as structural ensembles rather than isolated single conformations. The active and inactive states of molecular complexes can also be elucidated with these approaches.

Second, computational screening methodologies are likely to improve as well [10-12]. The fundamental physics of calculating free energies of interaction is well-understood. The primary barrier to high accuracy is the amount of processor time needed to carry out a full-scale simulation. With the expected continued increase in processor speed and parallel processing, the next decade should see simulation approach experimental accuracies. It is also reasonable to expect improved force fields and algorithmic advances as well. Of course there will always be cases of exceptional difficulty - large numbers of rotatable bonds, complex quantum interactions, coordinated proton transfer events. But we can expect ca. 0.1 kcals/mole accuracy in free energy calculations for many systems, with ca. 1 kcal/mole accuracy in enthalpy and entropy contributions to binding. An important issue for the pharmaceutical Industry is how to take part in these high-end calculations. Historically, major pharmaceutical companies would make the investment in the hardware. That might well continue to be the case because the overall cost is likely to the same as current computer hardware expenses. Alternatively, there will be ways to outsource these needs, especially for smaller companies.

Third, chemical synthesis strategies will change as computational and statistical methods make chemformatics more focused and more useful. For example, virtual libraries that organize the chemical expertise of the company and that exploit novel scaffolds will be easier to design. The advantages of "privileged" structures over *de novo* design will be much better understood.

Fourth, it is now feasible to calculate blood levels, cell permeability and first-order metabolic patterns for novel chemical species. At this writing, these methods are typically used to exclude parts of chemical space, but increasingly, there is predictive value in selecting lead series with relatively high marks in all the ADME areas rather than focusing first on affinity and later reworking the chemical structure to address delivery issues.

How targets are selected and validated will continue to evolve. Today's tendency to consider entire biological pathways must be the working paradigm. Modulation of the entire pathway will be the goal, rather than inhibition or enhancement of a single player. Further, each target will be seen in the context of all the related homologs and orthologs that could arise in, for example, the genetically diverse patient population, the preclinical animal species, and the infectious disease "genome". While the bookkeeping needed to make these data readily available is large, interested parties in academic and government-sponsored institutes will do much of the work. It is important for the pharmaceutical industry to consider how best to move these ideas forward. It will not be practical to have "secret" genomic information beyond the next 5-10 years.

There will clearly be enormous pressure on the process of choosing targets. It is easy to predict that single molecule "blockbusters" will become increasingly rare. One reason is that the genomic selection for clinical trials with its implications for labeling will reduce the size of the market for any one drug. Economic issues will also impinge, leading to increased concerns about efficacy and cost/benefit. At the drug development level, an effort will continue to quantify the "druggability" of targets [13]. Current attempts look at target sites using computational tools to assess the chances of coming up with tight-binding agents. A broader vision will also be in play: how druggable are different types of disease. There is a large amount of anecdotal comment on this later issue. We know that it is far easier to attack and "cure" infectious disease than degenerative disease. How is this insight to be quantitated? What types of strategies does it suggest at a corporate or industry-wide level?

Certainly, many challenges lie ahead in the scientific, medical, economic and political arenas. If history is helpful, we can recognize that much of the strength of the modern pharmaceutical industry rests on the promise to cure disease. This heritage from the successes of the antibiotic "wonder drugs" and the vaccines against polio and other childhood illnesses is still a powerful legacy that can guide our thinking. Can we revisit the standard that one drug treats more than one disease, and that we offer "cures" rather than temporary reprieves? This is an ambitious undertaking in these days of 10 or 12 medications for chronic aliments and multidrug resistance, but the greater understanding of basic biology and biochemistry offers great hope.

References

[1] N. Brooijmans and I.D. Kuntz. Molecular recognition and docking algorithms. *Annu. Rev. Biophys. Biomol. Struct.* **32** (2003) 335-373.

[2] I.D. Kuntz, J.M. Blaney, S.J. Oatley *et al.* A geometric approach to macromolecule-ligand interactions. *J. Mol. Biol.* **161** (1982) 269-288.

[3] W.L. Jorgensen. The many roles of computation in drug discovery. *Science* **303** (2004) 1813-1818.

[4] S. Shi and A. Kuki. A simple approach to a multi-objective design with constraints in compound selection for drug discovery. In: T.E. Simons (Eds): *Proceedings of the International Conference on Computational Methods in Sciences and Engineering*. Kastoria, Greece (2003) 587-592.

[5] S.B. Shuker, P.J. Hajduk, R.P. Meadows *et al.* Discovering high-affinity ligands for proteins: SAR by NMR. *Science* **274** (1996) 1531-1534.

[6] I.D. Kuntz, K. Chen, K.A. Sharp *et al.* The maximal affinity of ligands. *Proc. Natl. Acad. Sci. U.S.A.* **96** (1999) 9997-10002.

[7] A.L. Hopkins, C.R. Groom, A. Alex. Ligand efficiency: a useful metric for lead selection. *Drug Discov. Today* **9** (2004) 430-431.

[8] U. Pieper, N. Eswar, F.P. Davis et *al.* MODBASE: a database of annotated comparative protein structure models and associated resources. *Nucleic Acids Res.* **34** (2006) D291-D295.

[9] D. Devos, S. Dokudovskaya, R. Williams *et al.* Simple fold composition and modular architecture of the nuclear pore complex. *Proc. Natl. Acad. Sci. U.S.A.* **103** (2006) 2172-2177.

[10] J.C. Hermann, R. Marti-Arbona, A.A. Fedorov *et al.* Structure-based activity prediction for an enzyme of unknown function. *Nature* **448** (2007) 775-779.

[11] D.L. Mobley, A.P. Graves, J.D. Chodera *et al.* Predicting absolute ligand binding free energies to a simple model site. *J. Mol. Biol.* **371** (2007) 1118-1134.

[12] B.Y. Feng, A. Simeonov, A. Jadhav *et al.* A high-throughput screen for aggregation-based inhibition in a large compound library. *J. Med. Chem.* **50** (2007) 2385-2390.

[13] A.L. Hopkins and C.R. Groom. The druggable genome. *Nat. Rev. Drug Discov.* **1** (2002) 727-730.

Towards Drugs of the Future
C.G. Kruse and H. Timmerman (Eds.)
IOS Press, 2008

doi:10.3233/978-1-58603-949-3-29

Sense and Nonsense in Drug Discovery: A Chemical Perspective

Tudor I. Oprea
Division of Biocomputing, MSC11 6145, University of New Mexico School of Medicine, 2703 Frontier NE, Albuquerque, NM 87131, USA

Abstract. Drug discovery relies on our ability to mine the chemical and biological data in an integrated manner. From the initial choice of chemical descriptors to the similarity coefficient itself, each user- and software-made decision will modify the perception of chemical similarity and influence the results of chemical space exploration. Such decisions will alter even the meaning of "similar", leading to counter-intuitive neighbors. This effort is further complicated by the existence of chemical and target affinity cliffs, which is also noted in the area of marketed drugs. Indeed, similar molecules may have a different activity profile for the same array of properties, leading to a different clinical profile. Minor changes are sometimes responsible for the difference between "launched" and "withdrawn". The "Black Swan", a metaphor for the impact of the highly improbable, appears to be an inherent quality of the drug discovery enterprise. Predictions are based on the past, and can not substitute for proper design and use of experiments. Systems chemical biology may alleviate some of the inherent limitations related to prediction, by better integrating available experimental data.

Keywords. Cerivastatin, chemical space navigation, data mining, drug discovery, predictivity, similarity, systems chemical biology

1. Early Decisions in Chemical Space Exploration

Drug discovery within the context of pharmaceutical and biotech industry is a highly competitive arena, where appropriate and on-time decisions can significantly alter the chance of success of new products (drugs) on the market. The process rests on economic, clinical and scientific reasons, as marketing and clinical trial studies are used to decide the fate of drug candidates, in addition to scientific factors. In the scientific arena, as well as in the intellectual property (economic) aspects, cheminformatics plays a central role [1]. In this chapter we discuss our need to recognize the choices we make with respect to cheminformatics tools, the presence of rugged landscapes (cliffs), the possibility of failed predictions, and the need to provide better integration of experimental and theoretical tools.

The concept of "intramolecular diversity", illustrated for caffeine in Figure 1, suggests that a rather diverse set of approaches exist with respect to perception (e.g. aromaticity in SMILES) and, indeed, description (e.g. 2D or 3D, electrostatics or lipophilicity) of chemical structures. Thus, the initial choice of chemical descriptors will influence the neighborhood mapping in chemical exploration. The information overlap between 2D and 3D methods is likely to not exceed 40% [2], and is often related to size [3]. The formulas used to compute the similarity coefficients themselves, from cosine and Euclidian to Tanimoto and Mahalanobis, will further influence the neighborhood results. In fact, each user- and software-dependent decision is likely to influence the perception of chemical similarity, thus influence the results of chemical space exploration. Figure 2 illustrates such neighbors, when using simple properties such as molecular weight, MW, or the measured octanol/water partition coefficient, $LogP_{o/w}$ [4], to compare chemicals.

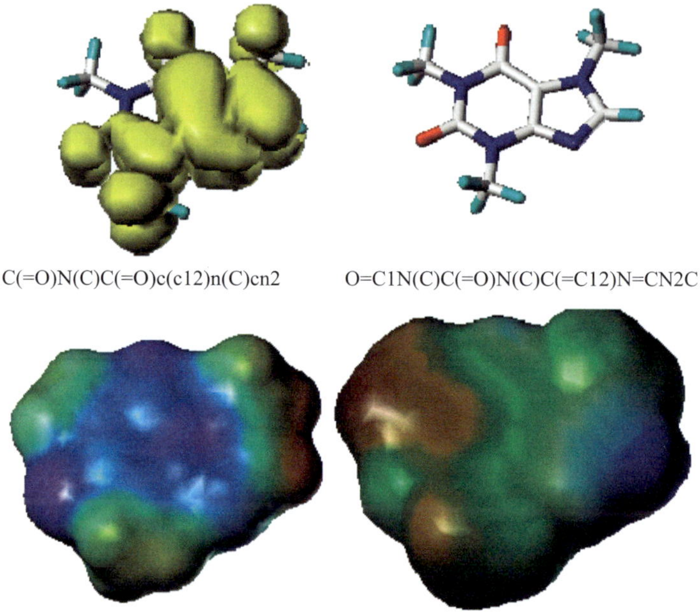

Figure 1. Chemical perception of caffeine: skeleton (stick) figure of the 3D structure (top right) is used for input for quantum chemical packages that calculate, e.g. the highest occupied molecular orbital (HOMO, top left), the molecular electrostatic potential at the Van der Waals surface (bottom left), or the molecular lipophilicity potential at the water accessible surface area (bottom right). SMILES codes [5] for caffeine illustrate the problem of uniqueness, as well as that of aromaticity perception, e.g. with respect to the 5-member ring of the xanthine scaffold. Modified from [6].

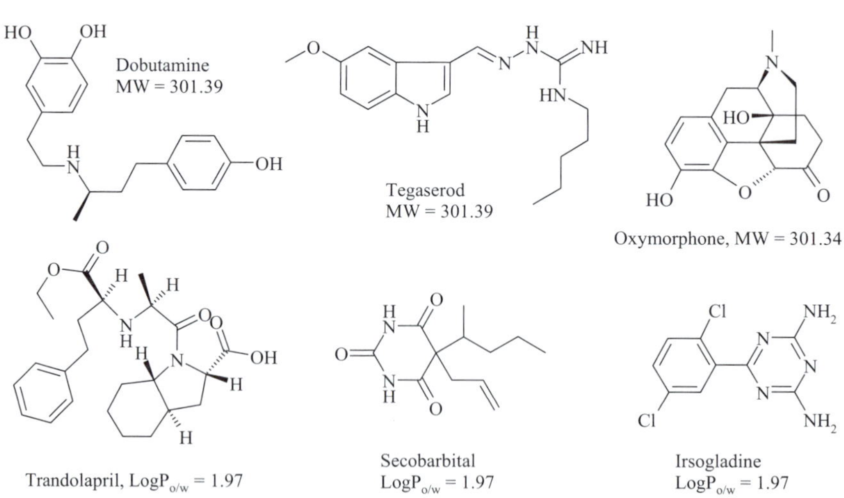

Figure 2. Launched drugs that share the same molecular weight (top) and $LogP_{o/w}$ (bottom).

Different metrics may result in unusual neighbors that belong to different chemical classes, as illustrated in Figure 2. Recognizing that such decisions influence the meaning of "similar" can be utilized constructively: our own virtual screening effort led to the identification of G-1, a novel non-steroidal G-protein coupled estrogen receptor (GPR30) agonist, starting from 17β-estradiol, by combining multiple similarity indices and descriptor metrics [7]. And yet, after almost three decades of property prediction, we continue to fail [8] on predicting $LogP_{o/w}$, a property deemed important by the vast majority of QSAR studies. Therefore, the notion of "similarity" needs to be understood in the context of not only the similarity index itself, but also the descriptor choice and implementation, and remains context-dependent. So, does the similarity principle hold in drug discovery space? Do similar drugs act on the same targets?

2. Affinity, Property and Market Activity Cliffs

Two decades ago, the tenet that "similar molecules act in a similar manner" was central to the process of drug discovery [9]. This approach forms the basis of analogue-based discovery [10], and has been validated through decades of empirical observations. However, as our ability to evaluate small molecules on a relatively large number of targets *in vitro* has expanded - process called "profiling", or "secondary pharmacology", we are beginning to realize that similar molecules can, in fact, have rather different profiles, or what Gerry Maggiora calls *activity cliffs* [11].

Table 1. Examples of compounds with multi-property activity cliffs. M1-M5, muscarinic receptors; %Oral - oral bioavailability; %PPB - plasma protein binding; Urine - fraction of drug excreted unchanged; T1/2 - plasma concentration half-life (hours); MRTD (μM) - maximum recommended therapeutic daily dose, in micromoles per kg-body weight per day. Data extracted from WOMBAT and WOMBAT-PK [12].

Chemical Structure	Property 1	Property 2	Property 3	Property 4	Property 5	Notes
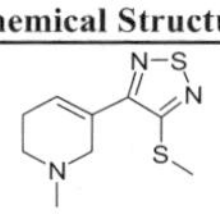	M1 IC50 = 8.34	M2 IC50 = 6.32	M3 IC50 = 5.8	M4 IC50 = 5.92	M5 IC50 = 5.68	
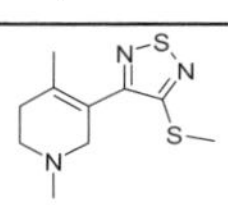	M1 IC50 < 6.3	M2 IC50 = 4.0	M3 IC50 < 4.0	M4 IC50 < 4.0	M5 IC50 < 4.0	
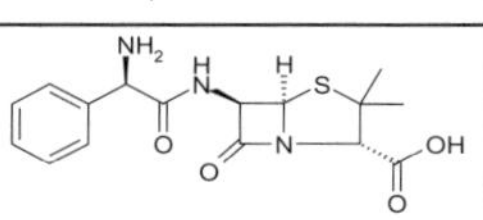	%Oral = 62	%PPB = 18	%Urine = 88	T1/2 = 1.3	MRTD (μM) = 166.85	Ampicilin
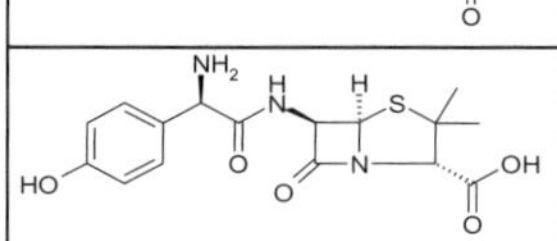	%Oral = 93	%PPB = 18	%Urine = 86	T1/2 = 1.7	MRTD (μM) = 68.41	Amoxicilin
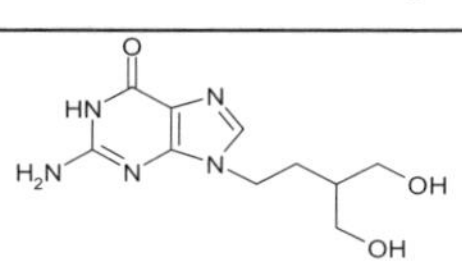	%Oral = 77	%PPB = 22.5	%Urine = 75	T1/2 = 10	Specific for the DNA Polymerase from HSV-1 and HSV-2	Penciclovir (active metabolite of famciclovir)
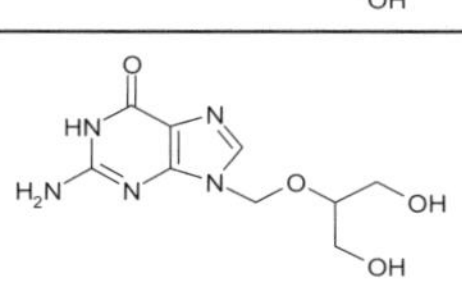	%Oral = 5	%PPB = 1.67	%Urine = 91	T1/2 = 3.5	Specific for the DNA Polymerase from CMV and HSV	Ganciclovir

As illustrated in Table 1, similar molecules can have quite different profiles for the same array of properties ("activity cliffs"). These differences range from biological activity towards the intended target (e.g. the muscarinic M1 receptor), or pharmacokinetic parameters (such as percent oral bioavailability), or indeed the observed clinical profile (e.g. the antiviral specificity of the acyclic nucleoside antiviral drugs). Thus, similar molecules may have a very different (clinical) profile, which influences *in vivo* properties such as efficacy, side effects, acute and chronic toxicity, etc. Therefore, the similarity principle is a good approximation for near-neighbors for a range of properties, but cannot be relied upon even within the narrow confines of the same chemical scaffold.

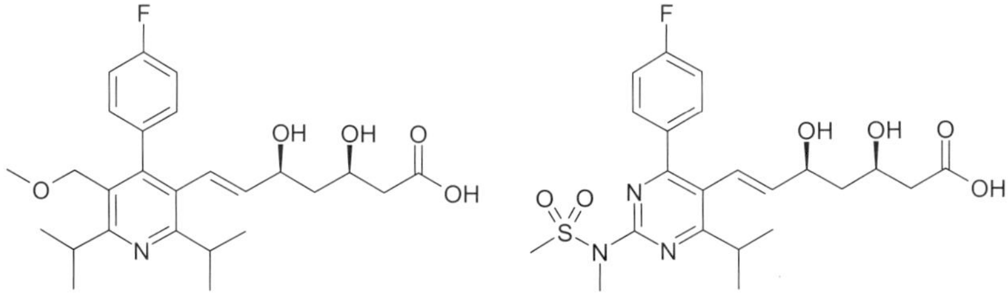

Figure 3. Cerivastatin (Baycol™, left) and Rosuvastatin (Crestor™, right); the common fragment is in blue.

Seemingly minor differences can be responsible for the "launched" or "withdrawn" status for drugs. Launched in 1997, Cerivastatin (Baycol™) was voluntarily withdrawn from all markets worldwide by Bayer in 2001, following reports of side effects of potentially fatal myopathy and rhabdomyolysis, in particular when the drug was co-administered with Gemfibrozil. At its peak, Baycol's global sales in 2000 exceeded $586 million USD. Launched by AstraZeneca in 2003, Rosuvastatin is now marketed in over 50 countries, with global sales over $2 billion USD in 2006. Both compounds are 3-hydroxy-3-methylglutaryl-coenzyme A (HMG-CoA) reductase inhibitors, and have been observed to have an antihypercholesterolemic and antihyperlipidemic effect by depleting cells of mevalonic acid, a cholesterol precursor. The event faced by Bayer in 2001, similar to the global withdrawal of Vioxx™ (Rofecoxib) by Merck and Co. in 2004 (global sales over $2.5 billion USD in 2003), can be categorized as a "Black Swan".

3. The "Black Swan": Sense and Nonsense in Discovery

Coined by Nassim Nicholas Taleb [13], the "Black Swan" is an unpredictable event, which has a massive impact from a business perspective. The term "Black Swan" is a metaphor for the first sighting of the black swan, which *(a)* invalidated the assumption that all swans are white - based on millions of previous observations; *(b)* changed our perception of those birds and *(c)* was retrospectively "assimilated" as a highly predictable event. Suggestively, Taleb conceptualizes the practical aspects of the "Black Swan" by depicting changes in the quality of life for a turkey, which steadily improves until meeting the butcher - an event that has no reference from the bird's perspective, or a "turkey surprise" [13]. Thus, the accumulation of past data cannot be used to predict the "unknown unknown".

Considered by philosophers and military alike, three cognitive categories can be discussed: *(a)* the "known knowns", for which sets and models are available, they have been validated externally and can be considered quite reliable; *(b)* the "known unknowns", for which sets and models can be documented and proved, but otherwise lack external validation (i.e. true predictivity is of limited reliability); and finally, *(c)* the "unknown unknowns", for which we do not have sets and models, thus lack any validation and predictivity. On the surface, the concept of the "unknown unknown" does not appear to be applicable for, e.g. Cerivastatin: good affinity for a known target (HMG-CoA reductase), well-categorized pharmacokinetic profile (e.g. 60% orally bioavailable, 80% bound to albumin), and documented clinical effect (it lowers cholesterolemia). The "unknown unknown" element was introduced by clinical practitioners, who prescribed this lipid lowering drug in combination with an older antilipidemic, Gemfibrozil (marketed since 1982 as Lipur™/Lopid™). Besides its serum lipid regulating effect, Gemfibrozil also blocks CYP2C8, the cytochrome P450 isozyme primarily responsible for metabolizing Cerivastatin. This unanticipated drug-drug interaction may have resulted in overdosing Cerivastatin, leading to 12 documented cases of fatal rhabdomyolisis. Since "the best study of man is man" (Alexander Pope), and since one cannot anticipate, during carefully controlled clinical studies, all the possible combinations of drug-drug and drug-food interactions, we argue that the process of drug discovery and the pharmaceutical industry are subject to "Black Swan" events [14].

Taleb further points the logical fallacy, *post hoc ergo propter hoc* that permeates in science: the appearance of correlation is often thought to relate to causality, i.e. a high fraction of adjusted variance ($R^2 \geq 0.7$) does not imply that two variables share a causality relationship. In an amusing example, the decline in the number of brooding stork pairs is suggested as being responsible for the falling birth rate in West Germany [15]. In drug discovery, however, common sense may not always be so easily exercised. We routinely exclude outliers (chemical; biological; statistical) from our datasets, in an effort to improve the R^2. When models fail, we consider that our choice of descriptors or perhaps the length of the molecular dynamics simulation time is inadequate. Some consider that the experimental data is incorrect - which is a distinct possibility; yet others might delete data when they do not fit the model! Outliers of any kind are rarely queried. This "unknown unknown" wealth of data is what most of us were trained to ignore. Yet serendipity has given us novel medicines, such as Penicillin and Viagra.

The crux of the matter remains our inability (à la "turkey surprise") to provide genuinely accurate predictions for unmapped regions of the chemical, biological and clinical space. For example, there is a wealth of literature on drug-likeness. Scientists (me included) have used "drug" as a quality to be studied with machine learning tools, i.e. computational models have been developed to discriminate "drugs" from other (non-drug) molecules. The quality of a drug is however attributed by a regulatory body, based on evidence available at the time of the IND application. Committees weigh the available evidence and decide to approve, and sometimes withdraw, the "drug" quality from the molecule in question. Thus, "drug" is not a natural property of chemicals and its models remain qualitative at best.

Predictions and, indeed, all models, are based on the past. The "known known" is first base, and is the only tool we have for exploring the "known unknown". Chemical and biological spaces can be mapped, but in incremental fashion. Within the limits of the applicability domain [16], and given an adequate set of descriptors, scientists have the ability to generate (wrong [17]) models that extend our frontiers of knowledge. Such models can rarely anticipate with accuracy what Maggiora calls "affinity cliffs" - the "turkey surprise" equivalent of medicinal chemistry space mapping. For example, we suggested one balances well-deployed diversity tools with random selections when assembling a chemical library for screening, with the explicit purpose to allow for the unexpected [18]. Such balanced choices need to be made in order to provide good coverage not only for the chemical space ("diversity"), but also for the biological space ("functional genomics"), in order to reliably map the "known unknown".

As reliable as these models are claimed to be within the constraints of the already-mapped (i.e. past) chemical space, one should not expect learning methods (machine or human) to perform well when predicting previously unmapped chemical and biological spaces, or the "unknown unknown". Scientists like to entertain the notion that tools to accurately predict the "unknown unknown" can be designed, perhaps using computers. Some like to point out that computers consistently outclass grand masters in chess competitions. They fail to recall that chess falls into the "known known" category - though large, the number of moves on the chess board is finite, thus knowable. Furthermore, our culture has a definite bias towards success stories, which are the overwhelming majority of accounts in peer-reviewed journals. Our collective understanding of the unanticipated, of dealing with failed hypotheses and ruined (or ruinous) models and experiments is limited to reading first-hand (auto)biographical accounts. When mapping the "unknown unknown", we are often deprived of learning how to deal with the outlier, the unpredictable, the "Black Swan". There is no substitute for experience, which in turn explains the lucrative business of scientific consultants.

Scientific breakthroughs are largely unplanned, and are more likely to occur when and where out-of-the-box thinking is encouraged. In the current climate of competitive pressure and "airplane assembly" analogies, management would have everyone believe that drug discovery is akin to engineering new planes. Hence, an increased emphasis is placed on execution, not creative thinking, and perhaps for good reason: drug discovery is financially more lucrative when performed by imitation ("me-too" drugs). Pharmaceutical industry analysts as well as managers ought to revisit these issues, since the currently compartmentalized R&D departments, e.g. by scientific discipline and therapy area, have been noted to suffer from an "innovation deficit" more than a decade ago [19]. Financial forecasts are critical to decision makers when considering the pursuit of small molecules in therapeutic markets. Yet, this sector too is subject to "Black Swans". One such example is the development of Omeprazole (Losec™) [20]: financial forecasts in the early eighties suggested the anti-ulceratives market is saturated by blockbusters Cimetidine (Tagamet™) and Ranitidine (Zantac™). This led to the apparently informed decision of closing down the Omeprazole project - decision that was overturned by scientists believing in its success. By 2001, at the peak of its sales, Omeprazole totaled over $6 billion USD worldwide in 2001; its *S*-isomer, esomeprazole (Nexium™), accumulated over $5.3 billion USD in 2006. As Taleb keeps pointing out [13], financial forecasting is subject to "Black Swans" as well, such predictions being valid only with severe limitations. So, can the predictive aspect of chemical and biological space mapping be improved? Do we make the most of the data?

4. Systems Chemical Biology

Recent years have witnessed significant increases in the amounts of data related to genes, proteins and their modulation by small molecules [21]. At the same time, modeling the regulation of entire genomes, metabolic or signal transduction pathways [22] is enabled by systems biology [23], an area that brings together theory, computational modeling, and high throughput experimentation. Systems chemical biology (SCB) is the field that integrates chemical knowledge into such biological databases [24], by encompassing currently available software with bio- and cheminformatic databases and biological network simulations. Although it appears as an incremental approach, SCB is likely to lead toward a systematic understanding of the way small molecules impact biological systems.

The variety of genomic, proteomic and small molecule databases amenable to integration is unprecedented. At this time, no single cheminformatic tool is poised to take advantage of this information. In one suggestion, SCB tools would permit chemical structures to be specifically described, visualized and modeled within system biology networks, would allow comparisons of the bioactivity properties of chemicals across all the available databases, and would enable relevant predictions of the effects of small molecules on biologic processes [19]. SCB tools could then become the engine for predicting the influence of putative perturbing small molecules on systems biology networks.

SCB tools would capture appropriate reactions and metabolite data to simulate metabolic pathways, using reaction kinetics data as input, and could further be adapted to simulate specific organisms to human data. Once realistic network models are generated, the influence of small molecules would then be evaluated using a combination of bioinformatic and cheminformatic technologies. Where protein structures are not known, bioinformatics can assist in generating homology models in preparation for virtual screening; by contrast, virtual screening, is more efficient when the target structure is supported by experimental data, such as from X-ray and NMR. SCB tools can then be used to query the impact of any novel small molecule on entire metabolic cycles, thus providing more realistic simulations of the impact of small molecules across biological pathways.

Thus, SCB tools can become a much more effective manner to map the chemical and biological space, perhaps providing more reliable methods to explore the "known unknown". The appropriate development of an integrated SCB interface is further likely to diminish the "unknown unknown" and the impact of the "Black Swan", as the total sum of genes, proteins and metabolites is ultimately knowable to the point of appropriately evaluating the influence of genetic variability, major food groups, xenobiotics and other environmental factors. Scientists will gain access to the "known" and "predictive" aspects of small molecule-biological network interactions, thus being able to more aptly evaluate the potential therapeutic impact of small molecule blockade of critical pathway steps. Ultimately, this will lead to a better understanding of why some, but not all, proteins within a pathway make druggable targets, and encourage an early focus on those proteins and protein-protein interactions that are the most likely to be of clinical relevance.

5. Conclusions

Drug discovery rests on economic, clinical and scientific reasons, which require cheminformatics and systems chemical biology. The process leading to selecting a successful candidate for launch is non-trivial, since similar molecules may exhibit a different profile, which influence efficacy, toxicity, etc. We argue that the similarity principle cannot be relied upon, as one is more likely to observe activity, target and marketing cliffs. Such large variations are likely to result in "Black Swans", leading to the withdrawal of drugs from the market, thus having serious financial consequences. The "Black Swan" is a metaphor for the impact of the highly improbable, and the influence of the "unknown unknown". The emergence of systems chemical biology may lead to a more effective manner to map the chemical and biological space, perhaps providing appropriate tools to diminish the impact of the "unknown unknown". Genetic variability, food-drug and xenobiotic-drug interactions are likely to be better understood within an integrative approach. This will result in a better way to focus on those that have clinical relevance.

Acknowledgment

This work was supported in parts by the National Institutes of Health grant U54 MH074425-01. Dr. Dan C. Fara (UNM) provided proofreading for this manuscript.

References

[1] T. Olsson and T.I. Oprea. Cheminformatics: a tool for decision-makers in drug discovery. *Curr. Opin. Drug Discov. Devel.* **4** (2001) 308-313.

[2] T.I. Oprea. On the information content of 2D and 3D descriptors for QSAR. *J. Braz. Chem. Soc.* **13** (2002) 811-815.

[3] T.I. Oprea. Chemical space navigation in lead discovery. *Curr. Opin. Chem. Biol.* **6** (2002) 384-389.

[4] A.J. Leo. Calculating LogPoct from structures. *Chem. Rev.* **4** (1993) 1281-1306.

[5] D. Weininger. SMILES, a chemical language and information system. 1. Introduction to methodology and encoding rules. *J. Chem. Inf. Comput. Sci.* **28** (1988) 31-36.

[6] T.I. Oprea (Eds). *Chemoinformatics in Drug Discovery.* Wiley-VCH, Weinheim (2005).

[7] C.G. Bologa, C.M. Revankar, S.M. Young *et al.* Virtual and biomolecular screening converge on a selective agonist for GPR30. *Nat. Chem. Biol.* **2** (2006) 207-212.

[8] Y.C. Martin. What works and what does not: Lessons from experience in a pharmaceutical company. *QSAR Comb. Sci.* **25** (2006) 1192-1200.

[9] P. Willett. *Similarity and Clustering in Chemical Information Systems.* Research Studies Press, Letchworth (1987).

[10] J. Fischer and C.R. Ganellin. *Analogue-based Drug Discovery.* Wiley-VCH, Weinheim (2006).

[11] G.M. Maggiora. On outliers and activity cliffs - why QSAR often disappoints. *J. Chem. Inf. Model.* **46** (2006) 1535.

[12] M. Olah, R. Rad, L. Ostopovici *et al.* WOMBAT and WOMBAT-PK: Bioactivity databases for lead and drug discovery. In: S.L. Schreiber, T.M. Kapoor, G. Wess (Eds.): *Chemical Biology: From Small Molecules to Systems Biology and Drug Design.* Wiley-VCH, Weinheim (2007) 760-786.

[13] N.N. Taleb. *The Black Swan: The Impact of the Highly Improbable.* Random House, New York (2007).

[14] T.I. Oprea and A.L. Hopkins. Black swans and white tablets. *Nature Rev. Drug Discov.* (submitted).

[15] H. Sies. A new parameter for sex education. *Nature* **332** (1988) 495.

[16] J. Jaworska, N. Nikolova-Jeliazkova, T. Aldenberg. QSAR applicabilty domain estimation by projection of the training set descriptor space: a review. *Altern. Lab. Anim.* **33** (2005) 445-459.

[17] George E.P. Box. "All models are wrong. Some are useful."

[18] M.M. Olah, C.G. Bologa, T.I. Oprea. Strategies for compound selection. *Curr. Drug Discov. Technol.* **1** (2004) 211-220.

[19] J. Drews. Innovation deficit revisited: reflections on the productivity of pharmaceutical R&D. *Drug Discov. Today* **3** (1998) 491-494.

[20] P. Lindberg. Omeprazole. In: J.B. Taylor and D.J. Triggle (Eds.): *Comprehensive Medicinal Chemistry II (volume 8).* Elsevier, Amsterdam (2007) 213-225.

[21] T.I. Oprea and A. Tropsha. Target, chemical and bioactivity databases - integration is key. *Drug Discov. Today: Technologies* **3** (2006) 357-365.

[22] E. Voit, A.R. Neves, H. Santos. The intricate side of systems biology. *Proc. Natl. Acad. Sci. U.S.A.* **103** (2006) 9452-9457.

[23] D.B. Kell. Theodor Bücher Lecture. Metabolomics, modelling and machine learning in systems biology - towards an understanding of the languages of cells. *FEBS J.* **273** (2006) 873-894.

[24] T.I. Oprea, A. Tropsha, J.L. Faulon *et al.* Systems chemical biology. *Nat. Chem. Biol.* **3** (2007) 447-450.

Towards Drugs of the Future
C.G. Kruse and H. Timmerman (Eds.)
IOS Press, 2008

doi:10.3233/978-1-58603-949-3-37

Synthesis at the Interface of Chemistry and Biology: How and Why?

Harry C.J. Ottenheijm
Department of Pharmacology and Toxicology, Cardiovascular Research Institute, University of Maastricht, P.O. Box 616, 6200 MD Maastricht, The Netherlands

Abstract. Medicinal chemistry is a biology-based and chemistry driven discipline. One of the main drivers in medicinal chemistry is synthetic chemistry.

We have seen great contributions to the process of drug discovery by synthetic chemistry. One might recall the era of natural product chemistry and of combinatorial chemistry. Nevertheless, one who has followed the developments in medicinal chemistry over the last decades gets the impression that the discipline is pushing against a wall without really breaking through.

The real need for this breakthrough is highlighted by the attrition rates: for nearly one third of newly assessed targets we are not able to find leads and - even more disturbing - one third of the leads can not be optimalized into early clinical candidates.

The burning question is how to bridge the gap between the structure of a compound and its properties. Which compound should be tested and which should be made next? For decades, medicinal chemistry has been linked to these questions.

Recent efforts are directed towards the making of properties rather than the making of chemical structures. The physical chemical properties to be made are being described in such a way that they can be handled by chemists. These developments will be discussed. One of them is the renaissance of natural products. The unique chemical space occupied by natural products has led to a renewed interest in them.

Finally, we will touch briefly upon a major disconnection between academia and pharmaceutical industry. In these two domains medicinal chemistry and in particular synthesis seem to have a different meaning. Solving this disconnection might bring closer an answer to the burning questions in medicinal chemistry.

Keywords. Medicinal chemistry, organic synthesis, combinatorial chemistry, natural products, drug properties

Synthesis

The boundaries of medicinal chemistry are not sharply defined nor are they constant. One discipline that has been constantly poaching on medicinal chemistry is synthetic chemistry. In the beginning of the twentieth century synthetic chemistry began to revolutionize the science of drug discovery. New synthetic drugs resulted, such as the barbiturates, anaesthetics, arsenical compounds, the sulphonamides and the penicillins.

During those decades nature was a constant source of fascinating chemical substances, many of which had important pharmacological properties. They provided unique research opportunities not only because of these properties but also because of their three-dimensional architecture. For instance, attempts to construct progesterone (Figure 1) in the period between 1930-1940 provided insights into carbon-carbon bond formation and oxidation reactions, and importantly set the stage for conformational analysis. This research finally led to the birth control pill, now used by millions of women around the world.

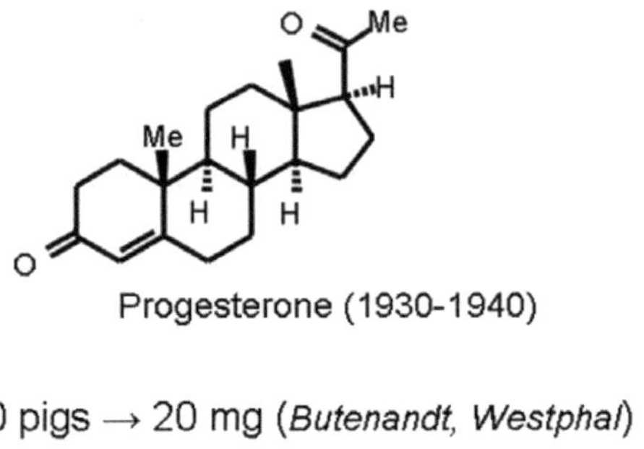

Figure 1.

The Process of Drug Discovery

In the period 1950-1980 the drug discovery process featured an interplay of medicinal chemistry and pharmacology. Compounds were designed and synthesized in gram quantities and tested in whole animals. These syntheses were time-consuming and the output was about 1-2 compounds per chemist per month. The pharmacologist then was able to handle and test this number of compounds. Thus a considerable amount of time used to pass (several weeks) before it was known whether a change in a molecule was a desired transformation.

The advent of new technologies, such as high throughput *in vitro* screening and defined molecular targets enabled the assaying of compounds at a much greater speed. To restore the equilibrium between synthesis and testing the chemists were challenged to speed up their synthesis programs. Automated synthesis and rapid separations were the result, and the term combinatorial chemistry was coined. The cycle time for synthetic manipulation of hits and leads decreased dramatically allowing for a much better-informed set of decisions when considering the next round of molecules to be prepared.

Combinatorial Chemistry

The advent of combinatorial chemistry about 17 years ago created huge excitement and many pharmaceutical industries quickly changed their drug discovery strategies in order to include several aspects of combinatorial chemistry. The basic premise was that it would generate libraries comprising millions of compounds to be screened by HTS and that it would produce drug leads by sheer number crunching.

Disappointingly, by the mid-1990s it became clear that the libraries created by then were not as useful as was hoped for. They yielded not the desired number of hits and leads and for some targets - in particular those related to anti-infectants, immunosuppression, oncology and metabolic diseases - the number was close to zero. This failure was contributed to several factors, such as the lack of diversity, the wrong chemical space and the absence of "drug-likeness".

Reasons for Reflections

Combinatorial chemistry is a technology that is there to stay (vide infra).

However, it is not the tool that enables the medicinal chemist to break through the wall he has been pushing against for decades. This barrier envisioned is discussed here.

First, a burning question is how to bridge the gap in our understanding of the structure-activity relationship. Which compounds should be tested and made next? The structure language as we still use it today is not well-suited to reflect the properties of a compound. Looking at their structures it is hard to understand why morphin and Leu-enkephalin have a similar activity at their receptor site.

Second, the success rates in finding good leads are low. The target classes we can approach with some confidence of obtaining a lead are the GPCR class 1, some enzyme classes, ion channels and nuclear receptors [1]. The cumulative attrition rate for all targets is 40% in lead identification and 37% in lead optimization [2], two phases in drug research where medicinal chemistry plays a dominant role.

So, what is needed is a breakthrough in our ability to modulate those target classes that are chemically hard to tract.

Third, whereas synthetic chemistry is a mature science, a recent survey [3] uncovered gaps in the synthetic methods used in the discovery and manufacture of drugs. The majority of the chemistries used is more than fifty years old. Many transformations involve functional group modifying reactions whereas the molecular construction reactions amount to only 48%. Some of the reactions that are needed but not yet available are given in Figure 2.

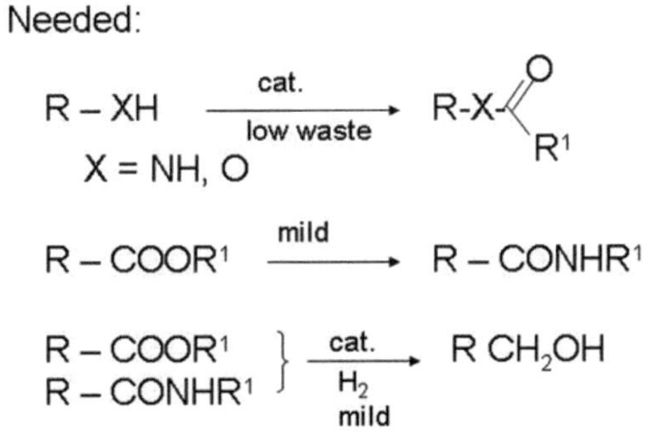

Figure 2. J.S. Carey *et al.* [3].

Some recent trends and developments addressing these three barriers are discussed here.

1. Which Properties Should Be Made?

A great challenge in medicinal chemistry is to gain insight into the bioavailability of potential drug candidates early in the drug discovery cycle. However, there is no consensus as to the physical properties relevant to the bioavailability. In Figure 3 some predictive rules are given that provide the probability of a bioavailability of a compound.

- **Pfizer's** rules: try to avoid:
  - •Molecular weight >500
  - •ClogP >5
  - •Number of H-bond donors >5
  - •Number of H-bond acceptors >10
- **Pharmacopeia's** rules: try to avoid:
  - •ClogP > 5.88
  - •PSA >131.6 Å^{2}
- **GSK's** rules: try to avoid:
  - •Rot. Bonds > 10
  - •PSA> 140 Å^{2}
- **Abbott's** ABS: try to avoid:
  - •PSA > 150 Å^{2} (neg. charged)
  - •Fail Pfizer rules (not negatively charged)

Figure 3.

Of those rules the one of Abbott deserves some attention [4]. Its essence is that the psychical chemical properties governing the bioavailability of a compound depend on its net charge. If a compound is neutral or positively charged, the rule of 5 (Lipinski's rule) can be applied. For anions the important property is polar surface area (see Figure 4). This rule holds for rat as well as human bioavailability.

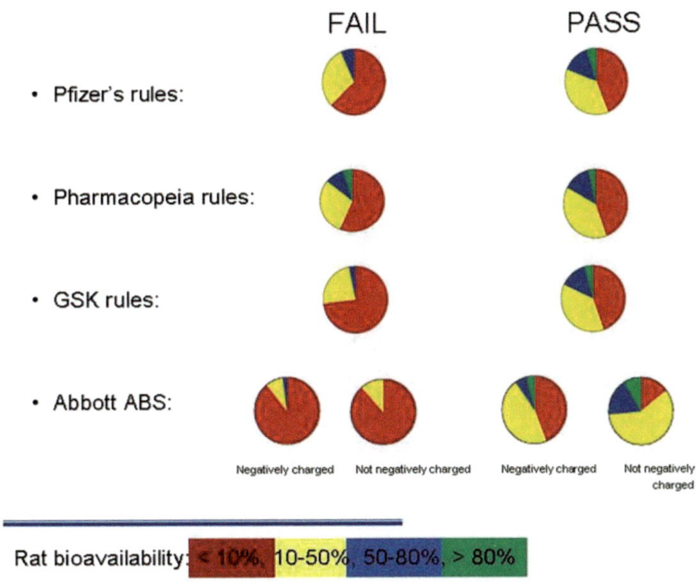

Figure 4. Y. Martin (Abbott) [4].

2. Attrition Rates

Results from early combinatorial libraries were often disappointing. However, in the last ten years more attention has been placed on the quality and diversity of combinatorial libraries. One recognizes the synergy between natural products chemistry and library construction based on combinatorial chemistry. Before we dwell upon this development, the impact of natural products chemistry on the process of drug discovery will be touched upon.

Natural Products as Leads for Drugs

Despite competition from other drug discovery methods, natural products have provided a substantial number of clinical candidates and drugs [5]. A fine example [5] is mevastatin (compactin), an antifungal agent from P.brevicompactum, that served as lead compound for a series of antilipidemic drugs, known as the statins (Figure 5).

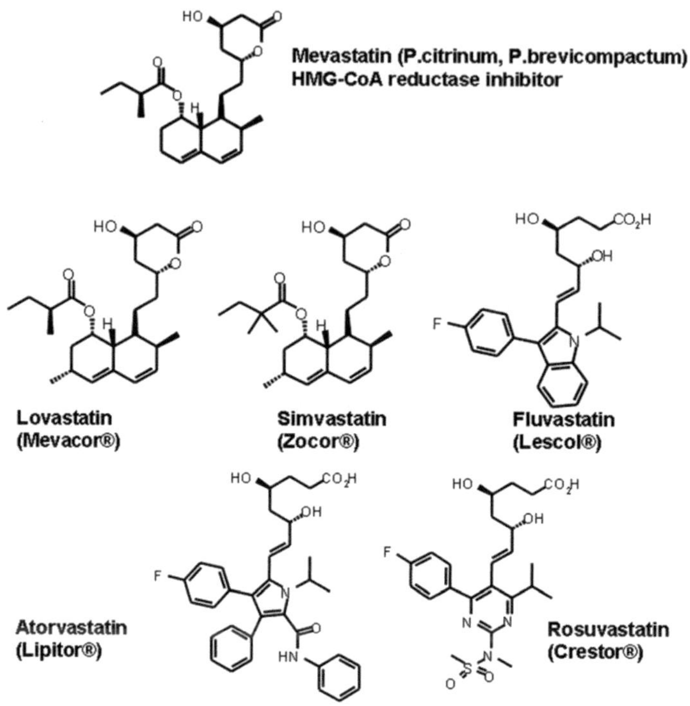

Figure 5.

In the totality of drug launches from the period 1981-2002 natural products were a significant source, especially in the anticancer and antihypertensive therapeutic areas [6]. Of all 29 small molecule drugs launched in 2000 eight were derived from natural products or hormones [7].

It is surprising therefore, that many pharmaceutical industries have scaled down their natural products research in the last 15 years. This is undoubtedly due to the advent of combinatorial chemistry. In an interesting development that diametrically opposed this scaling down, academia continued with (small molecule) natural products research and incorporated this in their research on combinatorial chemistry [8-14].

This development can be illustrated by the work on epothilones as antitumor drugs [14]. The four epothilones currently undergoing advanced clinical development are shown in Figure 6.

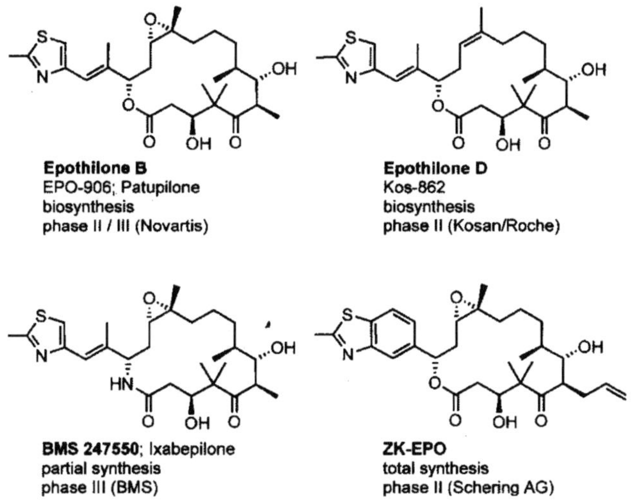

Figure 6. Epothilones in advanced clinical trials.

The first total syntheses of epothilone A were reported in 1996/97 by the groups of Danishefski, Nicolaou and Schinzer [15-17]. It was in particular Nicolaou who combined the traditional natural products synthesis approach with library generation. His groups applied the solid-phase, radio frequency-encoded chemistry, split-pool strategy. The synthesis elegantly features the olefin metathesis reaction to form and simultaneously cleave the macro-cycle from the solid support to yield four classes encompassing eight epothilones, see Figure 7 [9 and references cited therein].

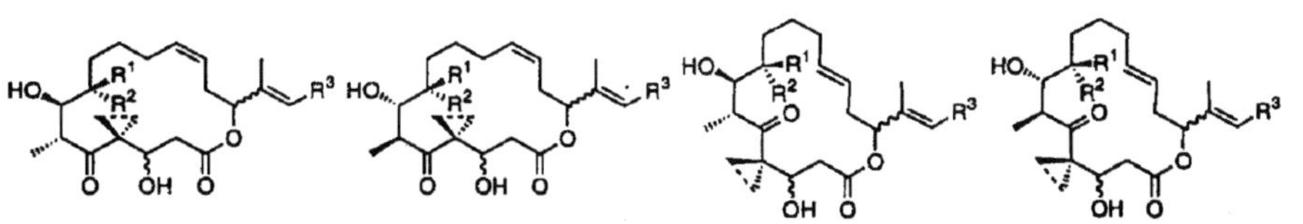

Figure 7. Synthesis of epothilone libraries by split-pool REC.

In the past decade it has become clear that small natural products occupy an important part of the chemical space that is formed by (potential) drugs. They are becoming an important subclass in the compound collections and serve as biologically validated starting points for library design. Recently, a structural classification of natural products was reported as organizing principle for charting the chemical space they occupy [18]. The aim of this classification is to arrange the natural products scaffolds in a tree-like fashion to provide a tool for the design of natural product derived compound collections.

One can conclude that natural products research has become more competitive with synthetic compound screening, due to the unique chemical space occupied by natural products.

Second Generation Combinatorial Chemistry

Due to the early disappointments in the results of combinatorial chemistry a reorientation has taken place in the last decade. Here follow some developments that fall within this reorientation. Striking is the coming together of combinatorial chemistry and natural chemistry.

Diversity-Oriented Synthesis

The traditional approach to natural products is target-oriented synthesis. A given structure is subjected to retrosynthetic analysis, identifying one or more simple starting materials, a process that goes from complex to simple. Reactions are then conceived in the forward sense towards the intended target structure.

Diametrically opposing this approval is diversity-orientated synthesis. The small molecule starting material used in this approach is subjected to reactions that aim at maximizing diversity. The starting material is selected on the basis of diversification using well-known reactions. Generation of rings by sequential, cascade, or tandem cyclisations is a common practice, often simulating biosynthetic processes for natural products. It was Schreiber in particular [19] who popularized this strategy that evolved from traditional solid-phase synthesis. The principle is depicted in Figure 8.

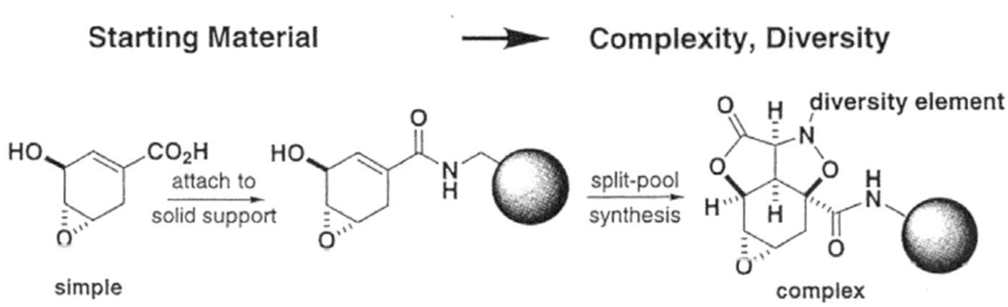

Figure 8.

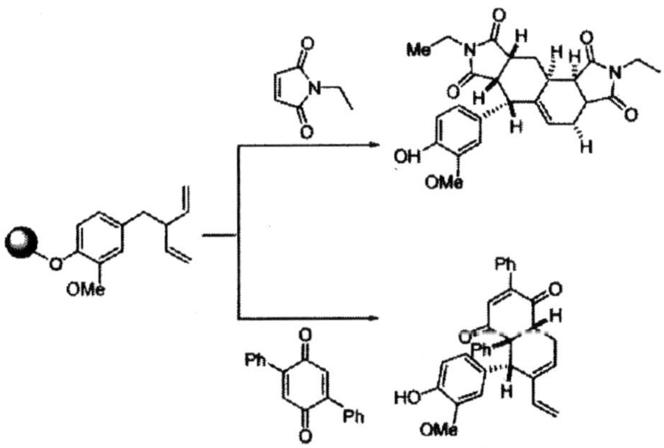

Figure 9. Representative example of differentiation DOS pathway leading to skeletal diversity.

Skeletal diversity can be obtained by employing starting material featuring pre-encoding skeletal information. The use of different reagents generates different scaffolds as illustrated in Figure 9 [20].

This approach seems well-suited for the finding of hits; one should bear in mind that the emerging molecules tend to have molecular weights higher than those generated by traditional combinatorial methods.

Target-Guided Ligand Assembly

There is great need for a strategy that can be employed when neither a hit nor structural or mechanistic information is available about the protein target. Such a strategy might prove to be target-guided ligand assembly. It features the following four steps [21], see also Figure 10:

a) Collecting a library of ligands that all posses a reactive group (X) capable of linking these ligands together.
b) Screening of the separate ligands and identification of binders, even weak ones.
c) Making a library combinatorial by linking together identified ligands; the linkers vary in length.
d) Finally, screening the resulting library.

Using this strategy a potent inhibitor of the tyrosine kinase c-Src was discovered, Figure 10 [21].

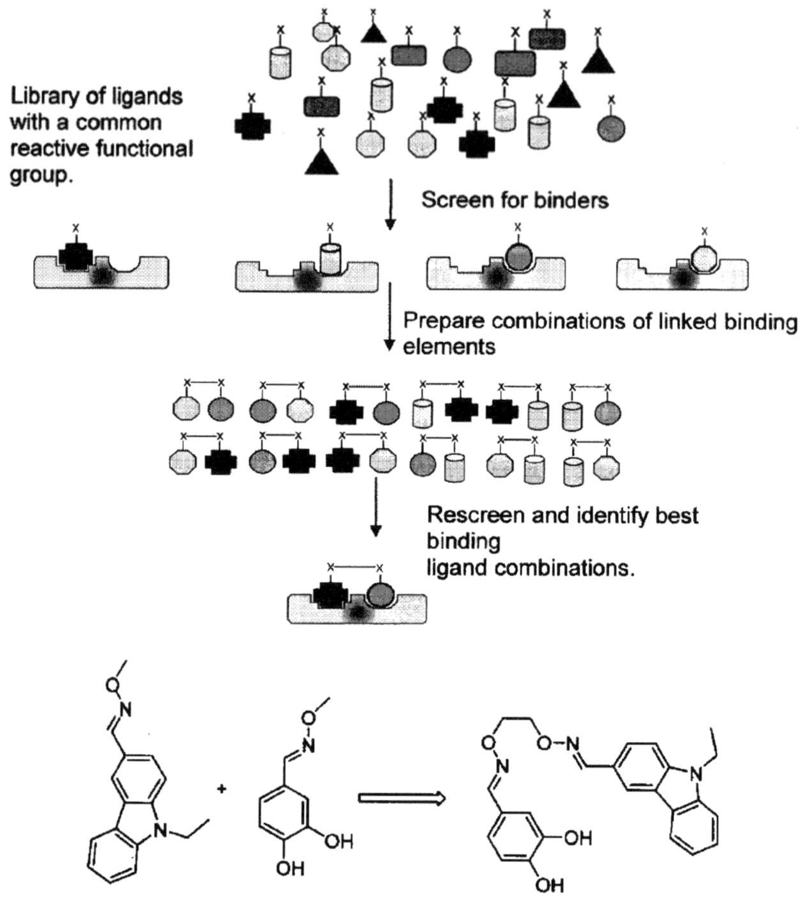

Figure 10. Assembly of a potent inhibitor from previously identified fragments, c-Src inhibitor with IC_{50} = 64 nM

Dynamic Combinatorial Libraries

So far we have seen the use of libraries comprising discrete molecules. The advent of supramolecular chemistry allows the creation of libraries featuring compounds capable of reversible interchange. In this approach, called dynamic combinatorial libraries stabilization of one member of the library affects the equilibrium (Figure 11), favoring the best binder [22,23]. Libraries made this way feature the principle of "self-replication and amplification" [23].

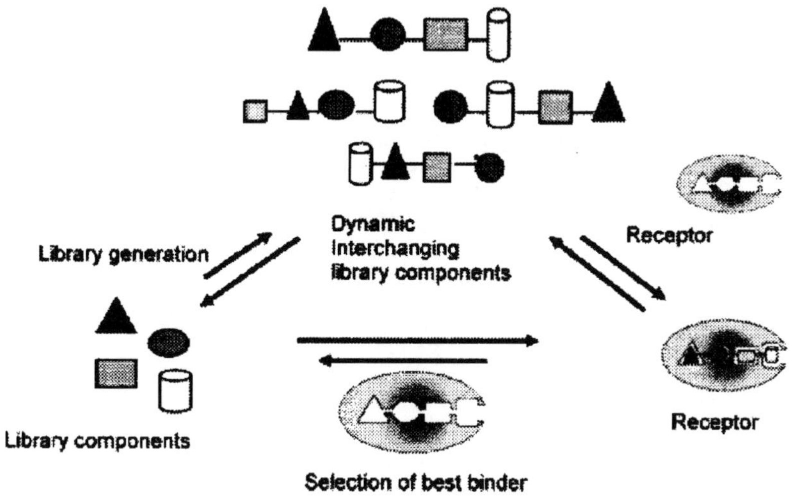

Figure 11. Dynamic combinatorial libraries.

This strategy has been used by J.M. Lehn [24] who identified an inhibitor of HPr kinase having a IC_{50}-value of 17 μM (Figure 12). Although the original lead had a similar IC_{50}-value (10 μM), this result nevertheless demonstrates the viability of this strategy.

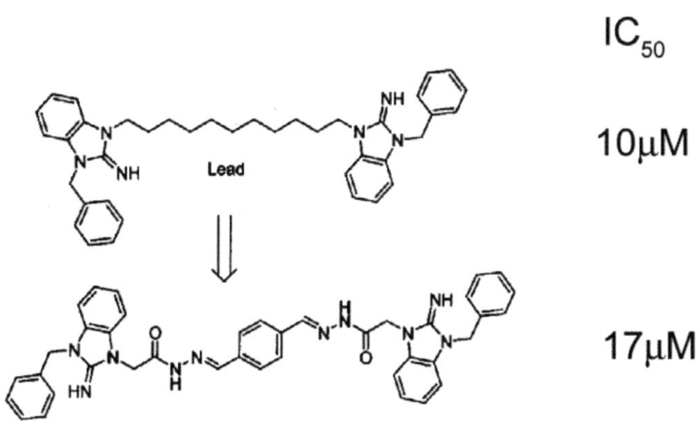

Figure 12. HPrK/P inhibitor by DCC.

3. Synthesis Methods

Biocatalysis

In view of atom economy there is a great need for chemical transformations on the basis of catalysis (see Figure 2). Metall catalysis and biocatalysis are already impacting chemical synthesis at the industrial scale. Here, biocatalysis is briefly touched upon.

It is emerging as a technology uniquely suited to the manufacture of drugs as a result of recent advances in DNA-sequencing, structural biology, protein expression, HTS and enzyme evolution technologies. These developments allow rapid identification of the desired biocatalyst. A fine example is the development of AG4704 [25,26] depicted in Figure 13.

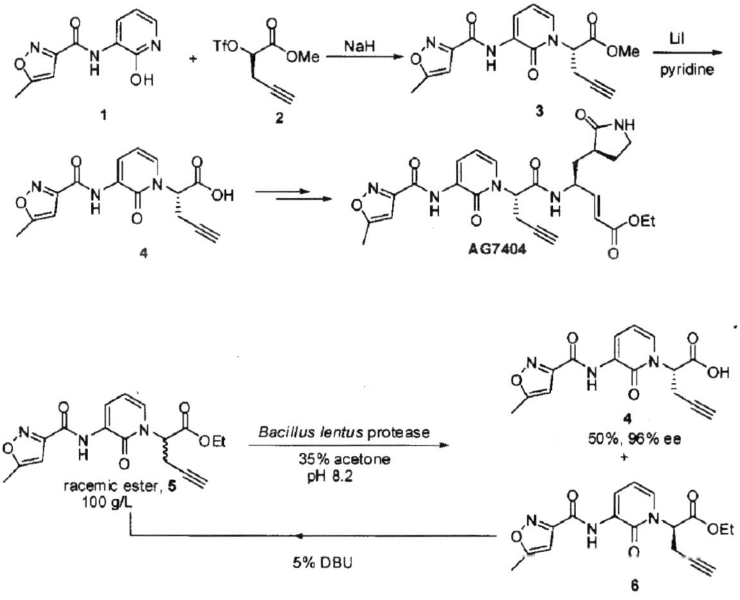

Figure 13.

In the preparation of this rhinovirus protease inhibitor, the key intermediate is the acid precursor 4, which originally was prepared from the triflate 2 and amide 1. The problem was that the homochiral triflate is not readily available and that the chiral center in 3 is prone to racemisation during the reaction conditions needed for the conversion of 3 into 4.

To resolve this issue an alternative chemo-enzymatic process was developed where the racemic ester 5 was converted into the homochiral acid 4 (50%, 90% ee). The wrong enantiomer 6 could be recycled into 5. The enzyme used was identified through a 96-well plate-based screening of a library of hydrolases. This enzyme had never been used before for synthetic applications.

It is obvious that this approach is of particular value at the industrial scale, where second-generation processes are needed to replace an existing chemical one.

Microwave-Assisted Synthesis

Microwave-assisted synthesis has rapidly gained acceptance as a valuable method for accelerating drug discovery and development [27]. The method is based on the ability of some liquids and solids to transform electromagnetic radiation into heat.

An example of the employment of this method in lead generation is depicted in Figure 14 [28]. A microwave promoted, three-component, one-pot reaction was developed to provide access to a tricyclic scaffold common to alkaloids with biological activities (yields 20-55%).

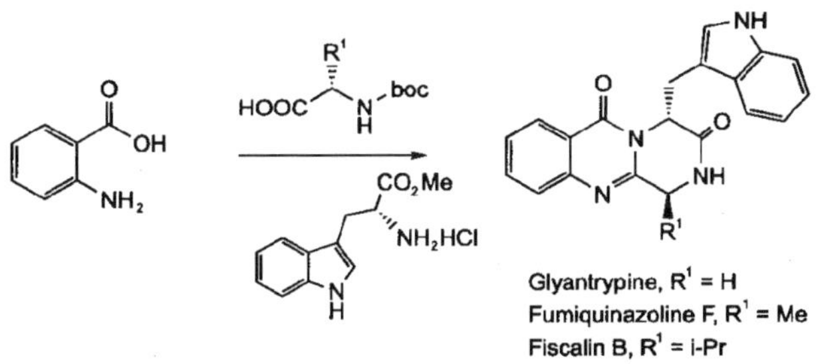

Figure 14. Example of multicomponent and flow-through microwave chemistry.

The Avoidance of Protecting Groups

Of the reactions used in the manufacture of drug candidates 21% involve protection-deprotection [3]. The complications imparted by protecting group manipulations have undoubtedly contributed to the notion that natural products were to be avoided in a drug discovery setting. To avoid the need for protecting groups an unorthodox approach is needed. Such an approach features making use of the intrinsic reactivity of a functional group rather than masking it.

Recently, such an approach towards a natural product, i.e. Ambiquine H, was reported [29]. Following standard procedures involving multiple protecting groups, others had prepared a precursor of the target compound in 20 steps (Figure 15). The novel approach required only 4 steps (20% yield) and 2 additional ones for the conversion of the same precursor into the final product (12% overall yield, gram scale). Rather than assuming that the reactivity of specific functional groups had to be shielded, the chemists harnessed this reactivity. They were inspired by studying the biosynthetic pathway. The similarity to the strategy used in diversity-oriented synthesis is apparent.

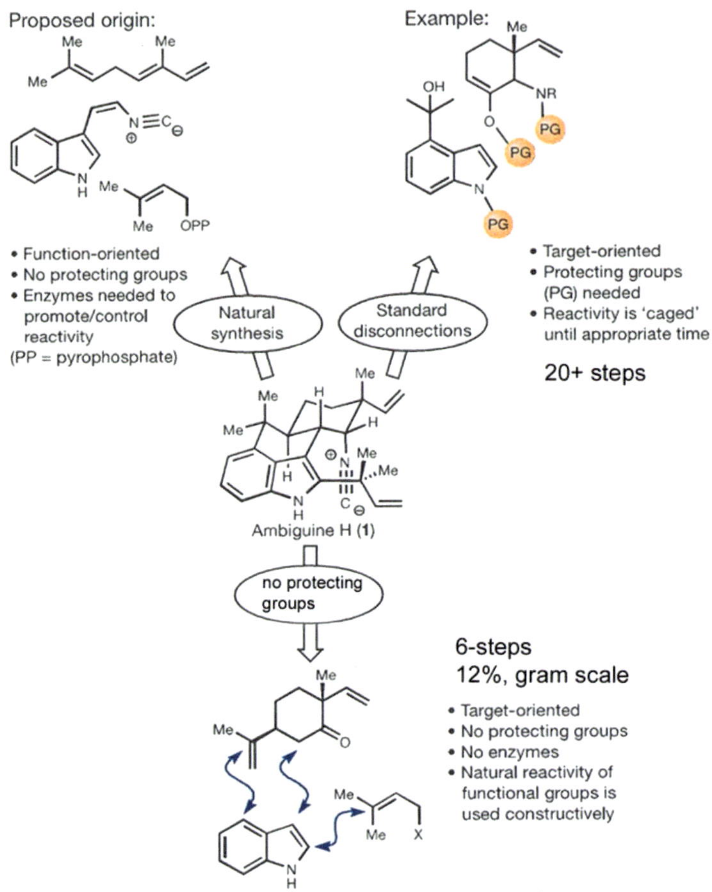

Figure 15.

Epilogue

In summary, some developments have been discussed that related to the question HOW? synthesis in medicinal chemistry is evolving. The interplay between natural products synthesis and combinatorial chemistry has been described.

Also the question WHY? has been addressed. Nobody less than George Hammond pointed out decades ago, that the goal of synthetic chemistry is the making of properties and not of structures. Meeting this challenge calls for further development of methods and strategies. It also requires improvement of the human factor, like chemical intuition, creativity and a prepared mind. It is there that teaching of medicinal chemistry has its place.

Teaching Medicinal Chemistry

There seems to be a disconnection between academia and pharmaceutical industry. In these two environments the term medicinal chemistry seems to have a different meaning, as has been pointed out by Ganellin *et al* [30]. Big pharmaceutical companies do not seek medicinal chemists for drug discovery. They hire instead scientist trained in synthetic organic chemistry, a number of which is metamorphosed into medicinal chemists by experience on the job. The adagio is "if you cannot make the compound, you cannot test it".

Academia as supplier train medicinal chemists broadly (synthesis, biochemistry, enzymology, molecular biology) but this seems less satisfactory to the customer, the pharmaceutical industry, since the strength in the training in synthesis was disconted, "synthesis cannot be learned effectively on the job".

One has no simple answer to this disturbingly apparent disconnection. However, bridging it will undoubtedly bring closer an answer to the burning question in medicinal chemistry: "How to prepare properties".

Acknowledgements

The author wants to thank F. Rutjes, C. van Boeckel, X. Bartoli, M. Timmers and D. Rees for making available some information used herein.

References

[1] P. Imming, C. Sinning, A. Meyer. Drugs, their targets and the nature and number of drug targets. *Nat. Rev. Drug Discov.* **5** (2006) 821-834. Erratum in: *Nat. Rev. Drug Discov.* **6** (2007) 126.

[2] D. Brown and G. Superti-Furga. Rediscovering the sweet spot in drug discovery. *Drug. Discov. Today* **8** (2003) 1067-1077. Comment in: *Drug Discov. Today* **9** (2004) 59-60; *Drug Discov. Today* **9** (2004) 381-383.

[3] J.S. Carey, D. Laffan, C. Thomson *et al.* Analysis of the reactions used for the preparation of drug candidate molecules. *Org. Biomol. Chem.* **4** (2006) 2337-2347.

[4] Y.C. Martin. A bioavailability score. *J. Med. Chem.* **48** (2005) 3164-3170.

[5] M.S. Butler. The role of natural product chemistry in drug discovery. *J. Nat. Prod.* **67** (2004) 2141-2153.

[6] D.J. Newman, G.M. Cragg, K.M. Snader. Natural products as sources of new drugs over the period 1981-2002. *J. Nat. Prod.* **66** (2003) 1022-1037.

[7] J.R. Proudfoot. Drugs, leads, and drug-likeness: an analysis of some recently launched drugs. *Bioorg. Med. Chem. Lett.* **12** (2002) 1647-1650.

[8] S. Hanessian. Structure-based organic synthesis of drug prototypes: a personal odyssey. *ChemMedChem.* **1** (2006) 1300-1330.

[9] K.C. Nicolaou, F. Roschangar, D. Vourloumis. Chemical biology of epothilones. *Angew. Chem. Int. Ed. Engl.* **37** (1998) 2014-2045.

[10] K.C. Nicolaou and S.A. Snyder. The essence of total synthesis. *Proc. Natl. Acad. Sci. U.S.A.* **101** (2004) 11929-11936.

[11] K.C. Nicolaou. Joys of molecules. 2. Endeavors in chemical biology and medicinal chemistry. *J. Med. Chem.* **48** (2005) 5613-5638.

[12] R.M. Wilson and S.J. Danishefsky. Small molecule natural products in the discovery of therapeutic agents: the synthesis connection. *J. Org. Chem.* **71** (2006) 8329-8351.

[13] A. Rivkin, F. Yoshimura, A.E. Gabarda *et al.* Complex target-oriented total synthesis in the drug discovery process: the discovery of a highly promising family of second generation epothilones. *J. Am. Chem. Soc.* **125** (2003) 2899-2901.

[14] U. Klar, B. Buchmann, W. Schwede *et al.* Total synthesis and antitumor activity of ZK-EPO: the first fully synthetic epothilone in clinical development. *Angew. Chem. Int. Ed. Engl.* **45** (2006) 7942-7948.

[15] A. Balog, D. Meng, T. Kamenecka *et al.* Total synthesis of (-)-Epothilone A. *Angew. Chem. Int. Ed. Engl.* **35** (1996) 2801-2803.

[16] Z. Yang, Y. He, D. Vourloumis *et al.* Total synthesis of Epothilone A: the Olefin metathesis approach. *Angew. Chem. Int. Ed. Engl.* **36** (1997) 166-168.

[17] D. Schinzer, A. Limberg, A. Bauer *et al.* Total Synthesis of (-)-Epothilone A. *Angew. Chem. Int. Ed. Engl.* **36** (1997) 523-524.

[18] M.A. Koch, A. Schuffenhauer, M. Scheck *et al.* Charting biologically relevant chemical space: a structural classification of natural products (SCONP). *Proc. Natl. Acad. Sci. U.S.A.* **102** (2005) 17272-17277.

[19] S.L. Schreiber. Target-oriented and diversity-oriented organic synthesis in drug discovery. *Science* **287** (2000) 1964-1969.

[20] M.D. Burke and S.L. Schreiber. A planning strategy for diversity-oriented synthesis. *Angew. Chem. Int. Ed. Engl.* **43** (2004) 46-58.

[21] D.J. Maly, I.C. Choong, J.A. Ellman. Combinatorial target-guided ligand assembly: identification of potent subtype-selective c-Src inhibitors. *Proc. Natl. Acad. Sci. U.S.A.* **97** (2000) 2419-2424.

[22] D.P. Walsh and Y.T. Chang. Chemical genetics. *Chem. Rev.* **106** (2006) 2476-2530.

[23] G.R. Cousins, S.A. Poulsen, J.K. Sanders. Molecular evolution: dynamic combinatorial libraries, autocatalytic networks and the quest for molecular function. *Curr. Opin. Chem. Biol.* **4** (2000) 270-279.

[24] T. Bunyapaiboonsri, H. Ramström, O. Ramström *et al.* Generation of bis-cationic heterocyclic inhibitors of Bacillus subtilis HPr kinase/phosphatase from a ditopic dynamic combinatorial library. *J. Med. Chem.* **46** (2003) 5803-5811.

[25] P.S. Dragovich, T.J. Prins, R. Zhou *et al.* Structure-based design, synthesis, and biological evaluation of irreversible human rhinovirus 3C protease inhibitors. 8. Pharmacological optimization of orally bioavailable 2-pyridone-containing peptidomimetics. *J. Med. Chem.* **46** (2003) 4572-4585.

[26] C.A. Martinez, D.R. Yazbeck, J. Tao. An efficient enzymatic preparation of rhinovirus protease inhibitor intermediates. *Tetrahedron* **60** (2004) 759-764.

[27] F. Mavandadi and A. Pilotti. The impact of microwave-assisted organic synthesis in drug discovery. *Drug. Discov. Today* **11** (2006) 165-174.

[28] J.F. Liu, P. Ye, B. Zhang *et al.* Three-component one-pot total syntheses of glyantrypine, fumiquinazoline F, and fiscalin B promoted by microwave irradiation. *J. Org. Chem.* **70** (2005) 6339-6345.

[29] P.S. Baran, T.J. Maimone. J.M. Richter. Total synthesis of marine natural products without using protecting groups. *Nature* **446** (2007) 404-408. Comment in: *Nature* **446** (2007) 383-385.

[30] C.R. Ganellin, L.A. Mitscher, B. Clement *et al.* University education of medicinal chemists: comparison of eight countries. *Eur. J. Med. Chem.* **35** (2000) 163-174.

Towards Drugs of the Future
C.G. Kruse and H. Timmerman (Eds.)
IOS Press, 2008

doi:10.3233/978-1-58603-949-3-53

Drug Design with ADME in Mind: Recent Paradigm Shifts in Drug Discovery

Ronald T. Borchardt
Department of Pharmaceutical Chemistry, 104 McCollum Research Laboratories, The University of Kansas, 2095 Constant Avenue, Lawrence, KS 66047, USA

Abstract. In an attempt to reduce the attrition rate of new chemical entities (NCEs) in development, many pharmaceutical companies have implemented a new drug discovery paradigm that involves optimizing the "pharmacological activity" as well as the "drug-like" properties (e.g. solubility, chemical/enzymatic stability, protein binding, cell permeability) during the lead optimization process in drug discovery. Traditionally, optimizing these "drug-like" properties was not considered by discovery scientists to be their responsibility. Instead, discovery scientists felt that undesirable "drug-like" properties of drug candidates would be "fixed" by preclinical development scientists. However, today most discovery scientists recognize that the "drug-like" properties of NCEs are intrinsic properties of the molecules and that it is their responsibility to optimize both the "pharmacological activity" and the "drug-like" properties of these molecules. In this chapter, the rational for this paradigm shift in drug discovery is discussed. In addition, a case history focusing on the optimization of the "drug-like" properties of cyclic prodrugs of an opioid peptide DADLE (H-Tyr-D-Ala-Gly-Phe-D-Leu-OH) is provided.

Keywords. Drug-like properties, pharmacological properties, ADMET/formulation properties, drug candidates, structural hits, structural leads, high affinity ligands, lead optimization, drug discovery and development interface, oral bioavailability, drug distribution into the brain

Introduction

Drug discovery and development is a very complex, costly, and time-consuming process [1,2]. Because of the uncertainties associated with predicting the pharmacological effects and the toxicity characteristics of new chemical entities (NCEs) in man, their clinical development is quite prone to failure [3-5]. In recent years, pharmaceutical companies have come under increasing pressure to introduce more rapidly new blockbuster drugs into the marketplace [1]. Companies have responded to these pressures by introducing new technologies and new strategies to expedite drug discovery and development.

The drug discovery and development process has traditionally been divided into three separate steps (i.e. discovery research, preclinical development, and clinical development) that should be integrated both organizationally and functionally. To be successful, each of these steps requires highly educated and experienced scientists, clinicians, and engineers who have great depth in their respective areas of expertise. While these individuals generally have significant depth in their areas of expertise, they often lack the scientific breadth and experience necessary to communicate effectively across disciplines [6]. The inability of these people to communicate across discipline lines, as well as the increasing size and complexity of pharmaceutical and biotechnology companies, has fostered the establishment of separate and distinct discovery research, preclinical development, and clinical development "silos" within these companies (Figure 1).

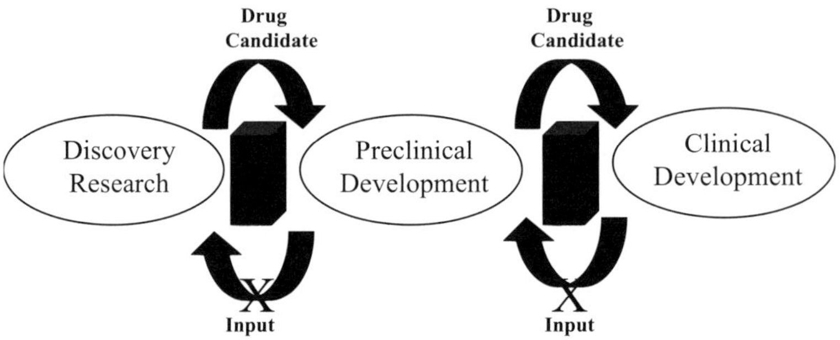

Figure 1. Drug discovery/development paradigm of the early 1990s.

Because of their isolation, scientists in the discovery research "silos" in pharmaceutical companies were in the 1980s and 1990s advancing an increasing number of "marginal" drug candidates into preclinical development [3-5]. These "marginal" drug candidates often lacked the characteristics needed to succeed in preclinical and clinical development. The increase in the number of "marginal" drug candidates being advanced into development arose in part because: (i) discovery scientists did not fully appreciate the "complete portfolio" of characteristics that a drug candidate must have to succeed in preclinical and clinical development; and (ii) a paradigm shift occurred in discovery research in the late 1980s and early 1990s that resulted in the generation of many drug candidates lacking "drug-like" properties (Figure 2).

ADME/Formulation Properties

Physicochemical

- Solubility
- Chemical stability
- Hydrophobicity/hydrogen bonding potential
- Charge
- Size
- Polymorphism

Biological

- Intestinal mucosal cell permeation
- Liver and kidney clearance
- Metabolism
- Transporters
- Protein binding
- Blood-brain barrier permeation
- Target cell permeation

Toxicity Properties

- Inhibition of HERG
- Inhibition of CYP450 isozymes
- PXR transactivation
- Human hepatocyte toxicity
- Mutagenicity
- Covalent protein binding

Figure 2. "Drug-like" properties.

This drug discovery paradigm shift involved a transition from using whole animal, disease-based screens to biochemical-based and cell-based screens during "lead" optimization (Figure 3). The whole animal, disease-based screens had the advantage that they afforded information about a molecule's "pharmacological" as well as its "drug-like" properties. In contrast, the biochemical-based and cell-based screens afforded only information about a molecule's potential "pharmacological" activity. This new drug discovery paradigm also saw the introduction of new biologies (e.g. genomics, proteomics, cell biology), new chemistries (e.g. combinatorial chemistry, informatics chemistry, parallel synthesis, structure-based drug design), high throughput screening (HTS), and bioinformatics into the drug discovery paradigm.

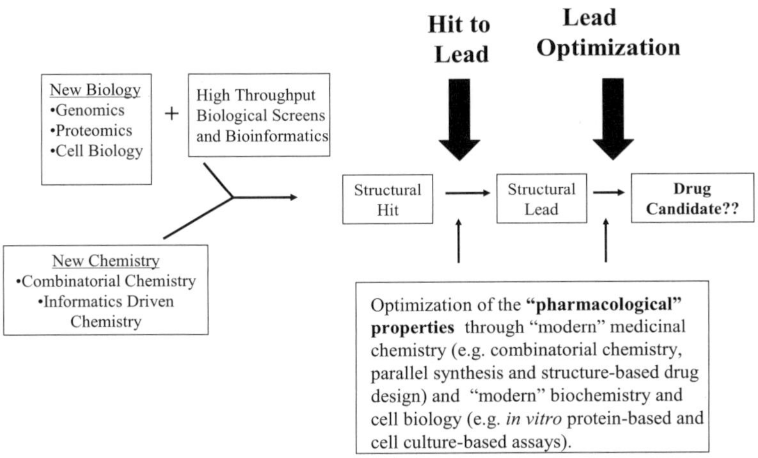

Figure 3. Drug discovery paradigm in the early 1990s.

Impact of the Biochemical and Cell Biology Emphasis in Drug Discovery on the Preclinical and Clinical Success of Drug Candidates

Unfortunately, this drug design paradigm that emphasized biochemical-based and cell-based screens tended to limit the participants in drug discovery to medicinal chemists and biologists (biochemists, cell biologists, molecular biologists, and pharmacologists) [6]. Input from preclinical development scientists, who have the knowledge and expertise in areas such as pharmaceutics, biopharmaceutics, pharmacokinetics, drug metabolism, toxicology, and process chemistry, into the "lead" optimization process was generally not sought out by medicinal chemists. The result was that scientists in discovery research were advancing what they thought were "high quality" drug candidates into preclinical development. However, in reality, these compounds were "marginal" drug candidates that might be better described as "high affinity" ligands. By definition, a drug candidate is a molecule that has high binding affinity and specificity for a validated therapeutic target, as well as "drug-like" properties (i.e. solubility, permeability, chemical/enzymatic stability, etc.). In contrast, a "high affinity" ligand is a molecule that has high binding affinity and specificity for a validated therapeutic target, but lacks certain "drug-like" properties. The absence of these "drug-like" properties could ultimately "lead" to the failure of these "high affinity" ligands in preclinical or clinical development or, at a minimum, make the development process more time-consuming and expensive.

Retrospective analysis data from the 1980s and 1990s has shown that compounds failed in preclinical and clinical development for various scientific reasons, including undesirable toxicity, lack of efficacy, and lack of optimal "drug-like" properties [3-5]. The failures arising from the lack of "drug-like" properties could be due to the paradigm shift in drug discovery described above and/or the availability of new knowledge and more precise and selective assays for characterizing a molecule's "drug-like" properties.

Traditionally, incorporating optimal "drug-like" properties into a structural "lead" was not considered by medicinal chemists to be their responsibility [7]. Instead, medicinal chemists felt that preclinical development scientists would "fix" the undesirable "drug-like" properties of their drug candidates. However, that view has changed in the past 10 years, resulting in another dramatic paradigm shift in drug discovery. The most significant aspect of this latest paradigm shift (Figure 4) is the recognition by medicinal chemists that the "drug-like" properties of structural "hits," structural "leads," and drug candidates are intrinsic properties of the molecules and that it is the responsibility of the medicinal chemist to optimize not only the "pharmacological" properties but also the "drug-like" properties of these molecules. Therefore, assessment of these "drug-like" properties is now done early in the drug discovery process (i.e. library design, profiling structural "hits" from HTS screens, transitioning structural "hits" to structural "leads," "lead" optimization, and candidate selection) (Figure 4) [8,9].

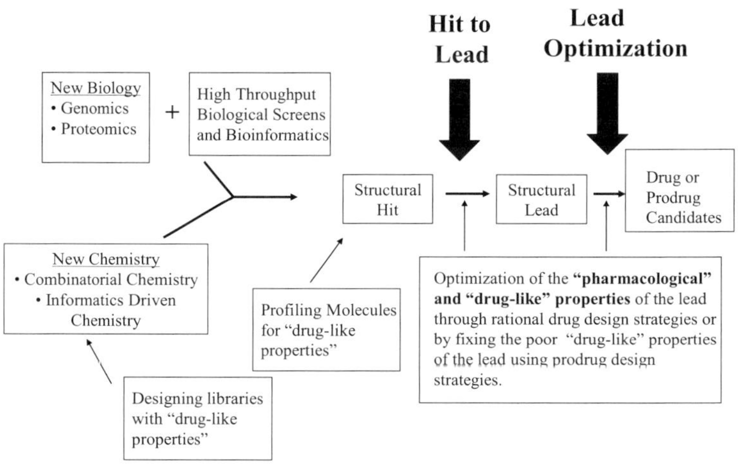

Figure 4. Current drug discovery paradigm.

Impact of "Drug-Like" Property Testing in Drug Discovery

Optimization of these "drug-like" properties (Figure 2) in the "hit-to-lead" stage (Figure 4) and the "lead" optimization stage (Figure 5) of drug discovery is now done through iterative processes in close collaboration with preclinical development scientists (Figure 6). These processes are analogous to those used by the medicinal chemist to optimize the "pharmacological activity" of a "hit" or a "lead." The implementation of this paradigm shift in drug discovery has created some people problems, technical problems, and strategy problems for pharmaceutical companies.

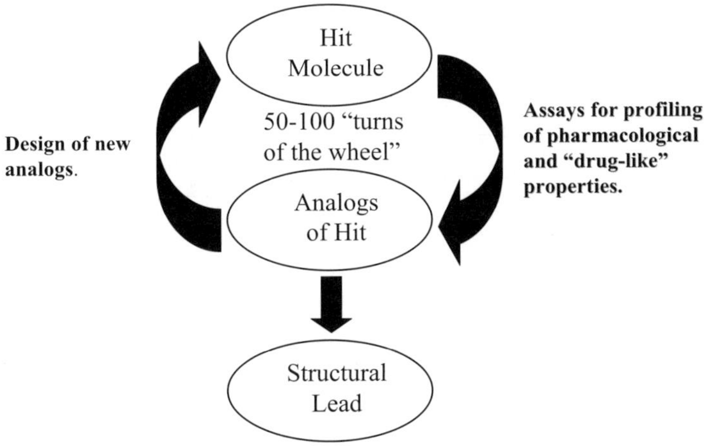

Figure 5. The iterative process is crucial to converting a hit to a lead in drug discovery.

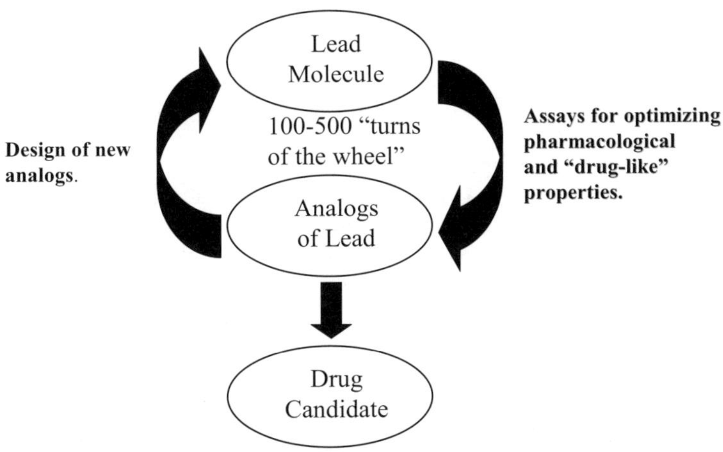

Figure 6. The iterative process is crucial to lead optimization in drug discovery.

People are one of the most crucial factors determining the success or failure of this new drug discovery paradigm within a pharmaceutical company [6]. As shown in Figure 7, this new drug discovery/development paradigm is seamless and highly integrated. This new paradigm has created an important new interface between what was traditionally the discovery research "silo" and the preclinical development "silo." Because scientists working at this drug discovery and development interface are expected to work closely with their colleagues in other disciplines, they need to have not only depth in their respective scientific disciplines but also the following skills: (i) scientific breadth in related disciplines; (ii) communication skills; (iii) interpersonal skills; and (iv) mutual respect for colleagues in other disciplines. In other words, these individuals need to have scientific breadth and also be team players. In many pharmaceutical companies, scientists working in the traditional discovery research "silo" and the preclinical development "silo" often lacked these skills. Therefore, companies need to encourage these individuals to expand their scientific knowledge and further refine their communication and interpersonal skills so that they can contribute more effectively to the implementation of this new drug discovery paradigm through their participation on multidisciplinary project teams.`

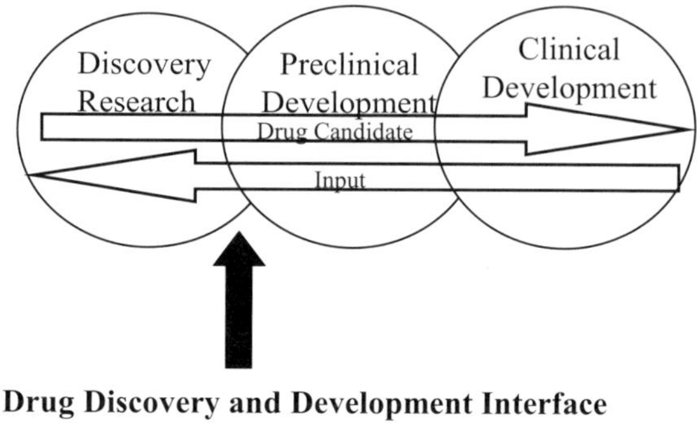

Figure 7. Today's drug discovery/development paradigm is seamless and highly integrated.

Because of a lack of scientific breadth, scientists in discovery research and preclinical development often have different scientific viewpoints [6]. These differences in viewpoints are due in part to a lack of appreciation and understanding by discovery scientists for preclinical and clinical drug development. Similarly, preclinical development scientists often lack an appreciation and understanding for the process of drug discovery. These differences in scientific perspective can be changed through the increased exposure of all the scientists, clinicians, and engineers in a company to the total process of drug discovery and development. Discovery scientists need to better understand and appreciate the highly regulated nature of preclinical and clinical development. Preclinical and clinical development scientists, clinicians, and engineers need to better understand and appreciate the qualitative and semi-quantitative nature of discovery research, the need for speed in the generation of "pharmacological" and "drug-like" experimental data during the "hit-to-lead" and "lead" optimization stages of drug discovery, and the difficulties and complexities of balancing the "pharmacological" and "drug-like" properties of drug candidates.

To successfully implement this new discovery paradigm, which focuses on simultaneous profiling of the "pharmacological" and "drug-like" properties during the "hit-to-lead" stage (Figure 4) and "lead" optimization stage (Figure 5) of drug discovery, new screening technologies needed to be developed [10-18]. These technologies had to be able to accommodate large numbers of compounds and rapidly generate experimental data. Compounds tested in the "hit-to-lead" step and "lead" optimization step of drug discovery are normally available in small quantities and they are ill-defined with respect to their physical form and purity (Figure 8). Because of the short timelines available to generate the data on the "drug-like" properties of "hits" and "leads" and the limitations on the quantity and purity of these compounds, the resulting data are often used to simply "bin" compounds under descriptors like "high," "medium" or "low." However, these qualitative and semi-quantitative measurements can be very helpful to guide discovery scientists in their efforts to optimize the "drug-like" properties of their "hits" and "leads" (Figure 9). Obviously, once "leads" are declared to be drug candidates, more quantitative studies would be done to guide the clinical development of these compounds (Figure 9).

Phases of Drug Discovery	Hit-to-Lead	Lead Optimization	Candidate Selection
# of Compounds	10^2-10^3	10^2	<3
Amounts	1-2 mg	5-15 mg	<1 kg
Physical Form	amorphous	amorphous/ crystalline	crystalline
Purity	<85%	>95%	>98%
Time	days	weeks	months

Figure 8. New technologies are needed to accommodate the increased number of compounds, reduced quantities of compounds, and the tighter timelines in drug discovery.

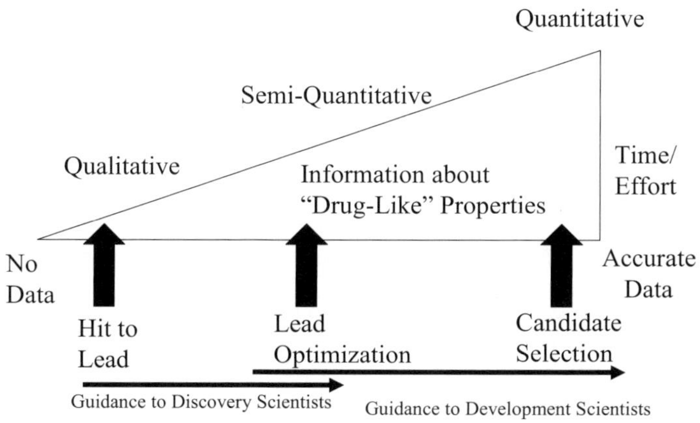

Figure 9. How extensively and at what stage of drug discovery should the "drug-like" properties of molecules be characterized?

To maximize the utility of the data forthcoming from "drug-like" property screening of "hits" and "leads," drug discovery project teams need to address the following question. **What "drug-like" property assays should be included in the discovery test funnels (Figure 10)?** Some project teams tend to answer this question by requesting data on as many "drug-like" properties as possible. The data arising from this strategy tend to confuse the discovery scientists because they often are unable to readily recognize which "drug-like" properties are most critical for the successful preclinical and clinical development of a drug candidate.

An alternative answer to the question posed above is first to ask a multidisciplinary project team to answer the following sub-questions:

- What is the anticipated route of administration (e.g. oral *vs.* i.v.)?
- What is the location of the therapeutic target (e.g. brain *vs.* heart)?
- What is the intended therapeutic indication (e.g. cancer *vs.* blood pressure)?
- What is the desired dosing regimen (e.g. b.i.d.)?
- What are the known undesirable "drug-like" properties (e.g. Pgp, HERG, 2D6, 3A4) of the structural "hits" and "leads"?

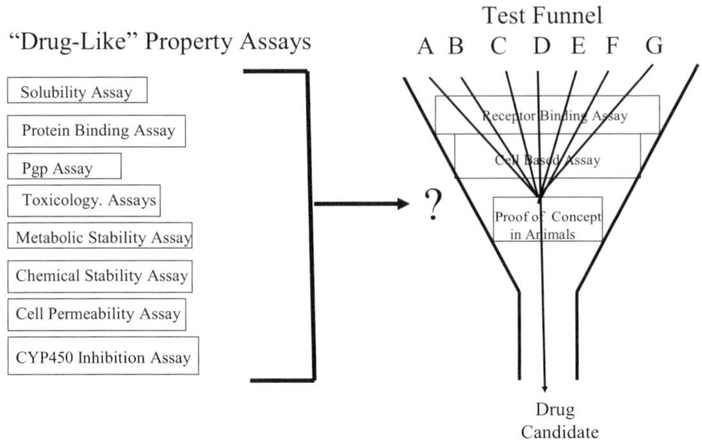

Figure 10. What "drug-like" property assays should be included in the test funnel and at what stage?

Based on the answers to these sub-questions, the project team can then design a rational testing strategy for both the "pharmacological" and the "drug-like" properties of the molecules synthesized during "hit-to-lead" and "lead" optimization [2,19-21]. However, project teams need to be flexible in what tests are included in the test funnel since medicinal chemists are good at fixing problems (e.g. poor permeability) but in the process often create new problems (e.g. poor solubility).

As an example, let's focus on an orally administered drug candidate targeting a receptor in a peripheral organ (e.g. heart). Figure 11 shows some of the key "drug-like" properties that will most impact the "developability" of such an orally administered drug candidate. In Figure 12 are shown some typical *in silico* and experimental determinations performed on molecules synthesized during the "hit-to-lead" stage of drug discovery [8]. Note that some of these experimental determinations are amenable to HTS and that the data arising from these determinations are often useful only for "binning" compounds (e.g. high, medium, and low solubility). In Figure 13 are shown some typical *in silico* and experimental determinations on molecules synthesized during the "lead" optimization stage of drug discovery [9]. These determinations tend to be done more rigorously from an analytical perspective; thus, the resulting experimental data are more accurate. These more accurate data are needed by medicinal chemists in order for them to detect subtle changes in the "drug-like" properties of compounds synthesized during "lead" optimization.

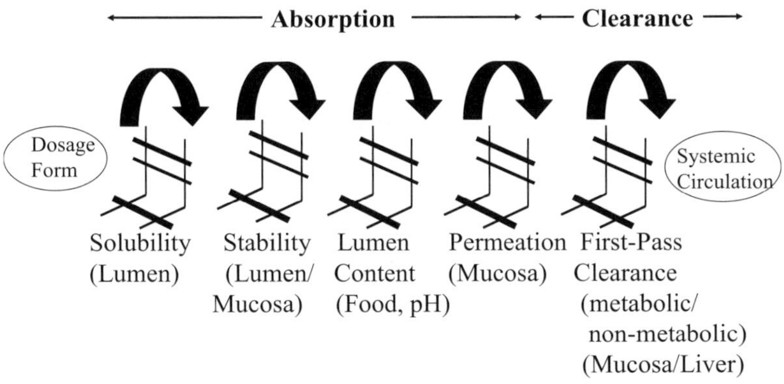

Figure 11. Key "drug-like" properties impacting "developability" of an orally administered drug candidate.

In silico

- Computational Predictions of Solubility, Permeability, Metabolism, Toxicity etc.

Experimental

- (HTS) Solubility (pH 6.5)
- (HTS) Chemical Stability (pH 1, pH 6)
- (HTS) Membrane Partitioning (PAMPA)
- (HTS) Cell Permeation (MDCK, Caco-2)
- (HTS) Microsomal/Hepatocyte Stability
- Vehicle selection for PK and Efficacy Studies
- Preliminary PK (including cassette dosing)

Figure 12. Profiling of ADME/formulation properties during the "Hit-to-Lead" stage of discovering a drug candidate for oral administration.

- Equilibrium Solubility (pH 1, pH 6.5)
- Mechanistic Cell Permeation Studies
- Stability in Microsomes/Hepatocytes (species comparisons)
- Profiling for Specific Cytochrome P450s
- Metabolite Identification
- Preliminary Physical Form Characterization
- Vehicle Selection for PK, Safety and Efficacy Studies
- Mechanistic Biopharmaceutical Studies
- Full PK Characterization
- *In Silico* Predictions of Oral Bioavailability

Figure 13. Profiling ADME/formulation properties during the "Lead" optimization stage of discovering a drug candidate for oral administration.

In addition to fixing poor "drug-like" properties using rational drug design strategies [8,9], medicinal chemists have in recent years embraced produgs as an alternative way to fix the poor "drug-like" properties of drug candidates [22,23]. Most commonly, prodrug strategies have been used to fix a drug candidate's poor permeability or poor solubility [24]. However, opportunities also exist to fix a drug candidate's poor chemical or enzymatic stability or to achieve organ targeting using prodrug strategies [24]. Similar to "lead" optimization, where the iterative process is crucial to optimizing both the "pharmacological" and "drug-like" properties, the iterative process is also crucial to optimizing the "drug-like" properties of a prodrug candidate (Figure 14).

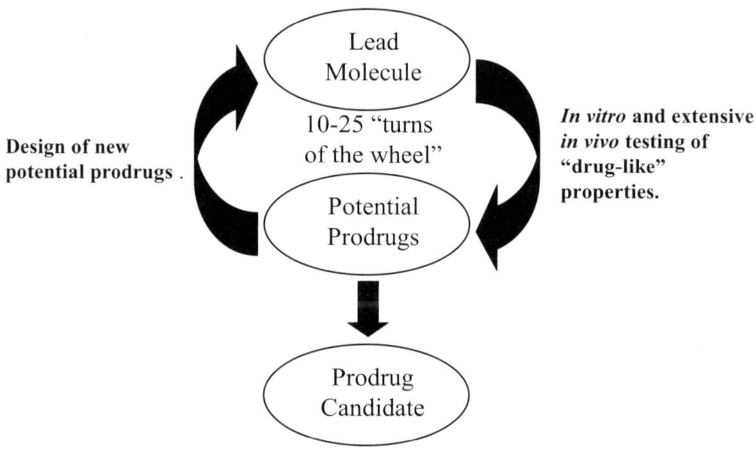

Figure 14. The iterative process is crucial to the discovery of prodrugs.

Case History: The DMPK Challenges of Designing Prodrugs of a Therapeutic Peptide that Are Orally Absorbed and Are Distributed into the Brain

To illustrate the importance of scientific breadth in the education of medicinal chemists, I have described below a research project conducted in my laboratory at The University of Kansas. The primary objective of this research project was to design prodrugs of DADLE, an opioid peptide (Figure 15), which would be orally absorbed and are distributed into the brain.

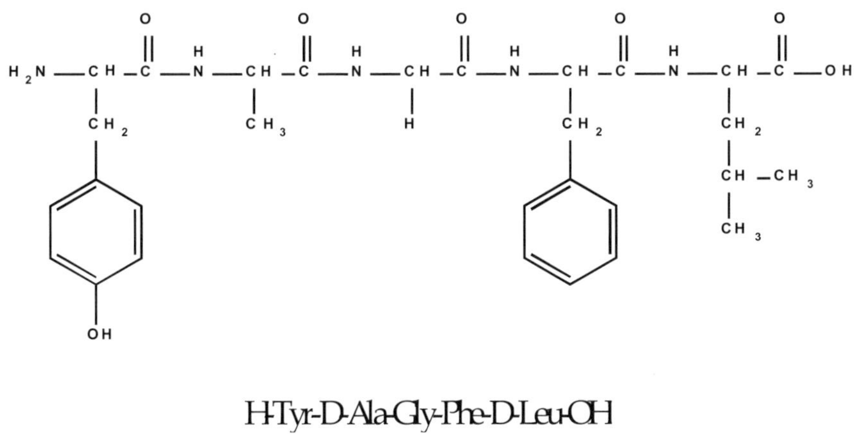

Figure 15. Structure of DADLE.

In describing this project, I have posed a series of questions that "modern" medicinal chemists should ask themselves as they design molecules to be synthesized and as they analyze data generated by DMPK scientists. While "modern" medicinal chemists do not need to conduct the DMPK experiments described below, they do need to be knowledgeable enough to suggest that these experiments should be conducted and they should be able to interpret the resulting experimental data. In addition, they should be able to use these data in the design of molecules to be synthesized during the "hit-to-lead" and "lead" optimization stages of drug discovery or prodrug discovery.

- **Based on the structural features of DADLE, would one expect this molecule to permeate across cell monolayers such as those found in the intestinal mucosa and the blood-brain barrier?**
  Based on empirical methods [11,12] and *in silico* methods [25], "modern" medicinal chemists would quickly conclude that DADLE would not permeate to any appreciable extent a cell monolayer. DADLE is a charged, hydrophilic molecule at physiological pH (Figure 16). While DADLE has some structural characteristics (charged and hydrophilic) found in typical solutes that permeate cell monolayers via the paracellular pathway, it has a relatively high molecular weight (570 DA). It is well-known that the permeability of a charged, hydrophilic solute via the paracellular pathway is highly dependent on its molecular weight (i.e. solutes having molecular weights of >300 DA exhibit very low paracellular permeation) [26,27].

  Since DADLE is a charged, hydrophilic molecule that contains numerous hydrogen-bonding acceptor and donors, "modern" medicinal chemists would predict that DADLE would exhibit very low permeation via the transcellular pathway. Support for this prediction was obtained using a chromatography method (i.e. immobilized artificial membrane (IAM) chromatography [28]) and a cell-based permeation assay (i.e. Caco-2 cell monolayer [29]). For example, the permeability coefficient (P_{app}) of DADLE in the apical (AP)-to-basolateral (BL) direction across a Caco-2 cell monolayer was shown to be extremely low (Table 1; 0.078×10^{-6} cm/s).

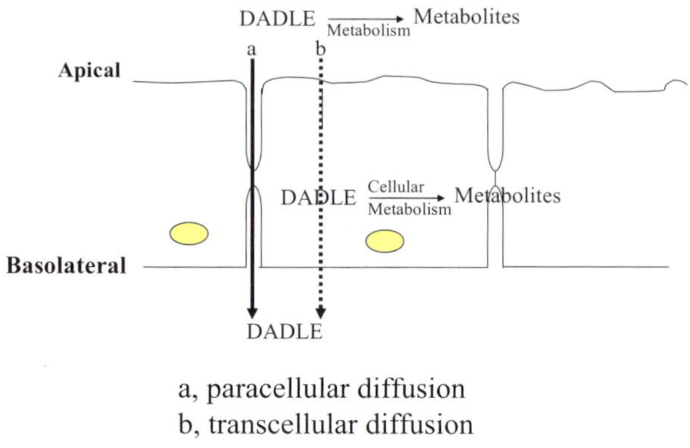

Figure 16. Possible pathways of DADLE permeation across the intestinal mucosa.

- **Can prodrug strategies be developed that would enhance the delivery of DADLE across cell monolayers like those found in the intestinal mucosa and the blood-brain barrier?**

  Being knowledgeable about prodrugs and how they have been used in the past to enhance the delivery of poorly permeable drugs [30], "modern" medicinal chemists would design prodrugs of DADLE that would be uncharged and lipophilic. Such prodrugs would be expected to have better permeability across cell monolayers via the transcellular pathway than DADLE.

  Our laboratory initiated such a prodrug design project in the mid-1990s. As shown in Figure 17, our strategy involved the design and synthesis of cyclic prodrugs of DADLE (e.g. AOA-DADLE, CA-DADLE, OMCA-DADLE) using an acyloxyalkoxy linker (AOA) [28], a coumarinic acid linker (CA) [31] or an oxymethycoumarinic acid linker (OMCA) [32]. These prodrugs were all shown to undergo bioconversion via a slow enzyme (esterase)-catalyzed hydrolysis of the ester bond formed from the carboxyl group of DADLE and the hydroxyl or phenolic group of the linker. This slow enzymatic step was followed by a fast chemical step [29,33,34], resulting in the formation of DADLE. This bioconversion mechanism is illustrated in Figure 18 using AOA-DADLE as an example.

  When designing prodrug strategies, "modern" medicinal chemists would realize that structural alterations made to change DADLE's cell permeation characteristics could lead to other changes in the molecule's "drug-like" properties (e.g. solubility, metabolic and non-metabolic clearance) [22,23]. The typical "drug-like" properties that a "modern" medicinal chemist would monitor during the optimization of an orally active prodrug of DADLE targeted to the brain are illustrated in Figures 19 and 20.

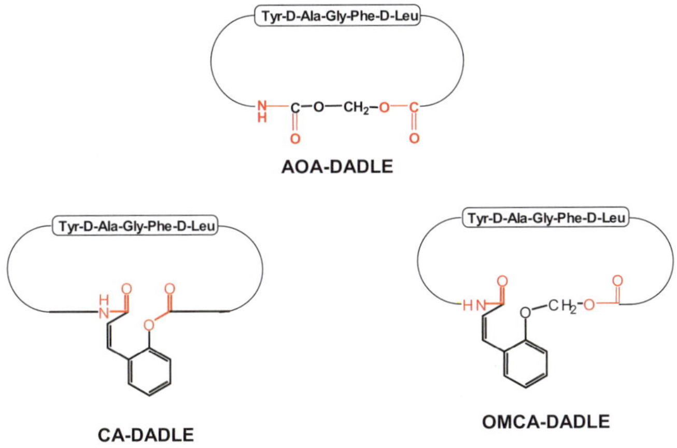

Figure 17. Cyclic prodrugs of DADLE.

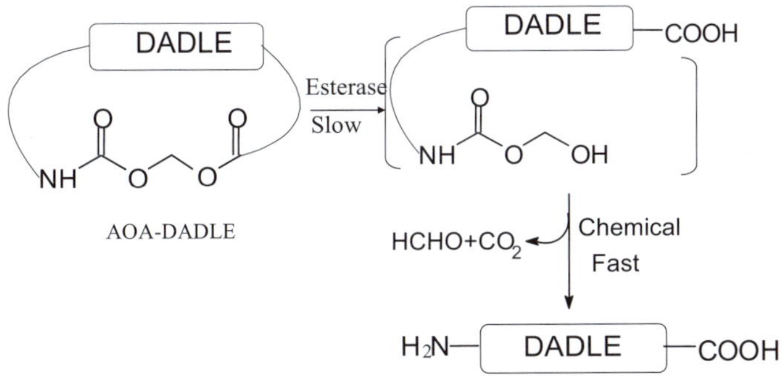

Figure 18. Bioconversion of the acyloxyalkoxy (AOA)-based cyclic prodrug of DADLE.

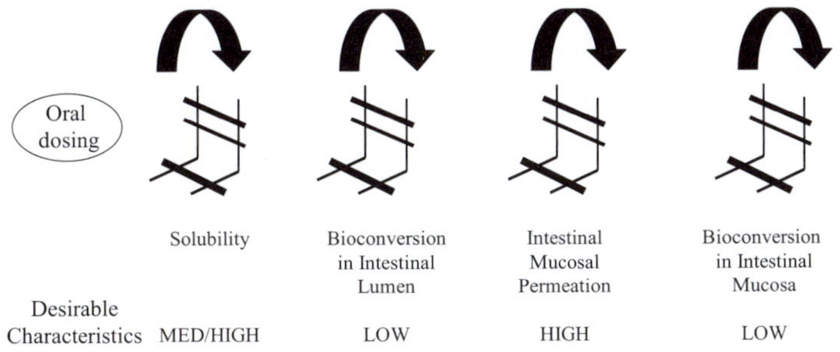

Figure 19. Key biopharmaceutical properties of an orally active prodrug targeted to the brain.

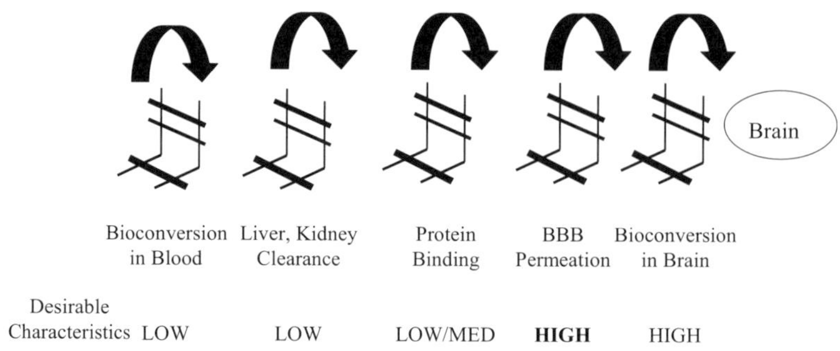

Figure 20. Key biopharmaceutical properties of an orally active prodrug targeted to the brain, *continued.*

- **Based on the structural features of AOA-DADLE, CA-DADLE and OMCA-DADLE, would one expect these cyclic prodrugs to have better transcellular permeation than DADLE across cell monolayers like those found in the intestinal mucosa and the blood-brain barrier?**
  Based on empirical methods [11,12] and *in silico* methods [25], "modern" medicinal chemists would quickly conclude that AOA-DADLE, CA-DADLE, and OMCA-DADLE have structural features (e.g. no charge, reduced hydrogen-bonding potential) typically found in transcellular permeants. The lack of charge arises from the prodrug design strategy that involved chemical modification of the N-terminal amino group and the C-terminal carboxyl group of DADLE using the linkers described above. The reduced hydrogen-bonding potential arises from the fact that these cyclic prodrugs can fold into unique solution structures (e.g. beta turns), which give rise to the formation of intramolecular hydrogen bonds. Support for these predictions was obtained from a chromatography method (e.g. IAM chromatography), spectroscopic techniques (e.g. 2D-NMR, CD) and molecular dynamics methods [28,35].

  However, when the permeation of AOA-DADLE, CA-DADLE and OMCA-DADLE was determined using a Caco-2 cell monolayer [29,33,34], the results were totally unexpected. These results are illustrated in Table 1 where the AP-to-BL P_{app} values for DADLE and AOA-DADLE are provided. These findings were surprising because the physicochemical characteristics of AOA-DADLE [32,35,36] suggest that it should be much more permeable across a Caco-2 cell monolayer than DADLE. Instead, the data in Table 1 show that DADLE is approximately four times more permeable than AOA-DADLE [29].

Table 1. Comparison of the AP-to-BL permeability coefficients of DADLE and AOA-DADLE in Caco-2 cells.

	P_{app}* 10^{-6} cm/s	Relative difference in permeability
DADLE	0.078 ± 0.007	1
AOA-DADLE	0.0186 ± 0.009	0.23

"Modern" medicinal chemists would be surprised (and probably disappointed) by these experimental results. However, they would quickly hypothesize that AOA-DADLE is a substrate for efflux transporters (e.g. Pgp) that are restricting the transcellular permeability of this prodrug [15,37]. These medicinal chemists would immediately request that their DMPK colleagues determine AOA-DADLE's permeation across a Caco-2 cell monolayer in the BL-to-AP direction. The results of this experiment would prove or disprove their efflux transporter hypothesis [15,21]. Data from this type of experiment are show in Table 2 [29]. The fact that the BL-to-AP P_{app} value for AOA-DADLE is approximately 50 times greater than its AP-to-BL P_{app} value strongly supports the hypothesis that efflux transporters (e.g. Pgp) are restricting the transcellular permeability of this prodrug across Caco-2 cells. Similar results have been observed for CA-DADLE [33] and OMCA-DADLE [34]. Further proof for this efflux transporter hypothesis was obtained using Madin-Darby Canine Kidney (MDCK) cells transfected with human Pgp (MDR1) or multidrug resistance-associated protein 2 (MRP2) [34,38,39].

Table 2. Comparison of the AP-to-BL and BL-to-AP permeability coefficients of AOA-DADLE in Caco-2 cells.

Permeability Direction	P_{app}* 10^{-6} cm/s	Ratio to $P_{B\ to\ A}/P_{A\ to\ B}$
AP-to-BL	0.0186 ± 0.009	52
BL-to-AP	0.969 ± 0.05	

- **Based on our knowledge about the role of efflux transporters (e.g. Pgp) in the intestinal mucosa, would one have learned anything from these Caco-2 cell and MDCK cell permeation experiments that would help predict the oral absorption of these DADLE prodrugs in rats?**

From the Caco-2 cell permeation characteristics of AOA-DADLE, CA-DADLE and OMCA-DADLE, "modern" medicinal chemists would began to question what effects the substrate activities of these cyclic prodrugs for efflux transporters would have on their disposition *in vivo*. From the extensive literature [18,37,40-43] now available on the role of transporters in determining the disposition of drugs, medicinal chemists would realize that the Caco-2 cells results might have implications for the oral absorption, the first-pass metabolic and non-metabolic clearance, and the BBB permeation of AOA-DADLE, CA-DADLE and OMCA-DADLE.

With respect to the effect on oral absorption of these cyclic prodrugs of DADLE, "modern" medicinal chemists would realize that Caco-2 cell and MDCK cell permeation data suggest that AOA-DADLE, CA-DADLE and OMCA-DADLE "might" exhibit low oral absorption. Therefore, they would ask their DMPK colleagues to orally dose these prodrugs to animals or conduct *in situ* rat intestinal perfusion experiments [44].

In Table 3 are presented data from experiments using an *in situ* rat intestinal perfusion model designed to determine the permeability of DADLE, AOA-DADLE, CA-DADLE and OMCA-DADLE (Ouyang *et al*, unpublished data). These data clearly show that DADLE is approximately four times more permeable than AOA-DADLE, CA-DADLE or OMCA-DADLE. These experimental results obtained using the *in situ* rat intestinal perfusion model are very similar to the experimental results obtained using the *in vitro* Caco-2 model of the intestinal mucosa [29,33,34]. The role of efflux transporters in restricting the oral absorption of these cyclic prodrugs in this *in situ* rat intestinal perfusion model was confirmed by determining their permeation in the presence of a Pgp inhibitor (PSC-833). In the presence of PSC-833, the P_B values of the cyclic prodrugs were approximately 30-45 times higher than in the absence of this Pgp inhibitor. Further, these results showed that the "intrinsic" intestinal mucosal permeation of AOA-DADLE, CA-DADLE and OMCA-DADLE was approximately 7-8 times greater than the "intrinsic" intestinal mucosal permeation of DADLE. These data clearly show that efflux transporters are restricting the permeation of AOA-DADLE, CA-DADLE and OMCA-DADLE across the rat intestinal mucosal.

Table 3. Intestinal mucosal $P_B{}^a$ values of DADLE and its prodrugs in the absence and presence of a Pgp inhibitor (PSC-833, PCS).

Compound	$P_B \times 10^8$ (cm/sec)		
	+PSC	+PSC[b]	Relative Increase in P_B
DADLE	24.6 ± 6.4	--	--
AOA-DADLE	4.0 ± 1.7	162 ± 36	40.5
CA-DADLE	5.6 ± 2.5	177 ± 56	31.6
OMCA-DADLE	4.2 ± 1.0	189 ± 27	45.0

Results are mean ± SEM, n > 3.

a. Apparent permeability coefficient (P_B) is based on appearance of the prodrugs in the blood.
b. 10 μM.

- **Based on our knowledge about the role of efflux transporters (e.g. Pgp) in the blood-brain barrier, would one have learned anything from these MDCK-MDR1 cell permeation experiments that would help predict the BBB permeation of these DADLE prodrugs in rats?**

Based on their knowledge of the literature about the role of efflux transporters (e.g. Pgp) in the BBB [18,37,41-43], "modern" medicinal chemists would hypothesize, based on MDCK-MDR1 cell permeation data described above [34,38,39], that the BBB permeation of AOA-DADLE, CA-DADLE or OMCA-DADLE would be very limited. To prove or disprove their hypothesis, medicinal chemists would ask their DMPK colleagues to orally or intravenously dose these prodrugs to animals and to measure the brain levels of the prodrugs and DADLE [18]. Alternatively, they would request that *in situ* rat brain perfusion studies be done with these prodrugs [45].

In Table 4 are presented data from experiments using an *in situ* rat brain perfusion model to determine the permeation of DADLE, AOA-DADLE, CA-DADLE and OMCA-DADLE through the rat BBB [46]. These data clearly show that, in spite of the differences in their physicochemical properties, the BBB P_{app} values of DADLE, AOA-DADLE, CA-DADLE or OMCA-DADLE are almost identical. The role of efflux transporters (e.g. Pgp) in restricting the BBB permeation of these cyclic prodrugs in this *in situ* rat brain perfusion model was confirmed by determining their permeation in the presence of a Pgp inhibitor (GF120918). In the presence of GF120918, the BBB P_{app} values of the cyclic prodrugs were approximately 50-450 times higher than in the absence of this Pgp inhibitor. Further, these results showed that the "intrinsic" BBB permeation of AOA-DADLE, CA-DADLE and OMCA-DADLE was approximately 120-370 times greater than the "intrinsic" BBB permeation of DADLE. These data clearly show that efflux transporters are restricting the permeation of AOA-DADLE, CA-DADLE and OMCA-DADLE across the rat BBB.

In contrast to the rat data described above, we have recently observed that in guinea pig OMCA-DADLE is much more able to permeate across the BBB after intravenous administration of the prodrug (Table 5) [47]. These differences in BBB permeation between rat and guinea pig are most likely due to the different substrate specificities of the efflux transporters that limit OMCA-DADLE's permeation into the brain.

Table 4. BBB P_{app}[a] values for DADLE and its prodrugs in the absence and presence of a Pgp inhibitor (GF120918).

Compound	P_{app} x 10^7 (cm/sec)[b]		Relative Increase in P_{app}
	-GF120918	+GF120918[c]	
DADLE	0.5 ± 1.4	0.6 ± 0.14	--
AOA-DADLE	1.2 ± 1.0	60.5 ± 25.6	50
CA-DADLE	0.4 ± 0.7	185 ± 68.3	460
OMCA-DADLE	0.7 ± 0.6	119 ± 12.8	170

a. Apparent permeability coefficients (P_{app}) were calculated from measured K_{in} and P_A based on the rat brain capillary surface area as reported (130 cm^2/g).
b. In the case of AOA-DADLE, CA-DADLE, and OMCA-DADLE, P_{app} values are based on the sum of prodrug, intermediate and DADLE presented in brain tissue.
c. 10 μM GF120918 was used in these studies since this is a concentration that totally inhibited Pgp.

Table 5. Uptake of OMCA-DADLE and its metabolites into rat and guinea pig brain.

Species	Compound	Brain Uptake (ng/g tissue)	
		Prodrug	Linear Peptide[b]
Rat[a]	OMCA-DADLE	3.30 ± 0.33	1.34 ± 1.50
Guinea Pig	OMCA-DADLE	263.6 ± 114.8	11.5 ± 2.6

a. Yang *et al.* [48]
b. DADLE

- **Based on knowledge about the overlap in substrate specificity between Pgp and cytochrome P450-3A4, would one have learned anything from these MDCK-MDR1 permeation experiments that would help predict the metabolism of these DADLE prodrugs in rat liver?**
Based on their knowledge of the literature [40] about the similarities in the structure-activity relationships of substrates for Pgp and cytochrome P450-3A4, "modern" medicinal chemists would hypothesize, based on the MDCK-MDR1 cell permeation data described above [34,38,39], that AOA-DADLE, CA-DADLE and OMCA-DADLE would be metabolized *in vivo* by Phase I enzymes. To prove or disprove their hypothesis, medicinal chemists would ask their DMPK colleagues to conduct stability studies of these cyclic prodrugs using rat liver microsomes [21].

In Figure 21 are presented data from experiments using rat liver microsomes to determine the stability of DADLE, AOA-DADLE, CA-DADLE and OMCA-DADLE (Ouyang *et al*, unpublished data). These data clearly show that in contrast to DADLE, which is stable when incubated with rat liver microsomes, AOA-DADLE, CA-DADLE and OMCA-DADLE are extremely unstable (i.e. >90% metabolized in 30 minutes). Further, the metabolism of these cyclic prodrugs appears to be mediated primarily by oxidation catalyzed by cytochrome P450s. The evidence in support of this hypothesis comes from the observation that ketoconazole (a known inhibitor of CYP3A) but not para-oxon (a known inhibitor of esterases) can prevent the metabolism of AOA-DADLE, CA-DADLE and OMCA-DADLE. These results strongly suggest that the oral bioavailability of these cyclic prodrugs would be limited in part by first-pass metabolism in the intestinal mucosa and the liver mediated by cytochrome P450s. In addition, it can be concluded that the clearance of these cyclic prodrugs may be mediated in part by cytochrome P450 metabolism in the liver.

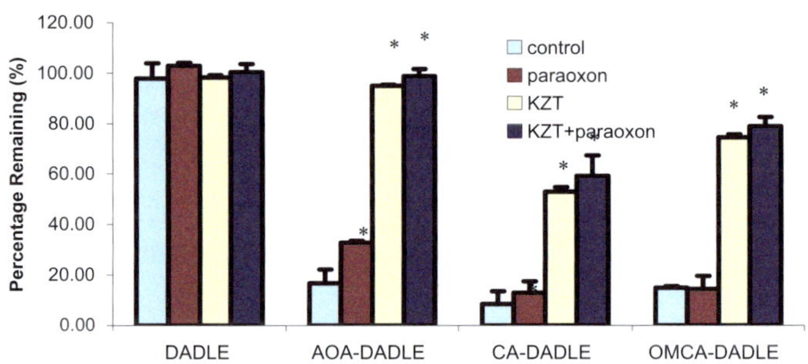

*: $P < 0.05$ compared to control (without inhibitor).
--- Paraoxon @ 100 μM; KZT @5 μM; prodrugs @ 2.5 μM incubated 30 min.

Figure 21. Stability of DADLE and its prodrugs in rat liver microsomes.

- **Based on our knowledge about the role of Pgp in the liver, would one have learned anything from these MDCK cell permeation experiments that would help predict the non-metabolic clearance of these DADLE prodrugs by rat liver?**
  Based on their knowledge of the literature about the role of efflux transporters (e.g. Pgp) in the liver [17,19,37,41-43], "modern" medicinal chemists would hypothesize, based on MDCK-MDR1 and MDCK-MRP2 cell permeation data described above [38,39], that AOA-DADLE, CA-DADLE and OMCA-DADLE could be cleared by non-metabolic mechanisms and eventually excreted in the bile. To prove or disprove this hypothesis, medicinal chemists would ask their drug metabolism colleagues to orally or intravenously dose these prodrugs to bile duct-cannulated rats and then measure the levels of the prodrugs and DADLE in bile.

  In Figure 22 are presented data from experiments designed to measure the extent to which AOA-DADLE, CA-DADLE and OMCA-DADLE undergo biliary (non-metabolic) clearance in rats [48]. In these experiments, the prodrugs were dosed to bile duct-cannulated rats and the bile was sampled at the indicated times for 2 h. The resulting data show that 45.5%, 25.7% and 38.1% of AOA-DADLE, CA-DADLE and OMCA-DADLE, respectively, were recovered in bile after 2 h [48]. Thus, in rats, these prodrugs undergo clearance via metabolic (Phase I metabolism) and non-metabolic (biliary clearance) pathways.

  In contrast to the rat data described above, we have recently observed that in guinea pig OMCA-DADLE undergoes very little biliary clearance after intravenous administration of the prodrug (Table 6) [47]. These differences in biliary clearance between rat and guinea pig are most likely due to the different substrate specificities of the efflux transporters that facilitate the liver clearance of OMCA-DADLE.

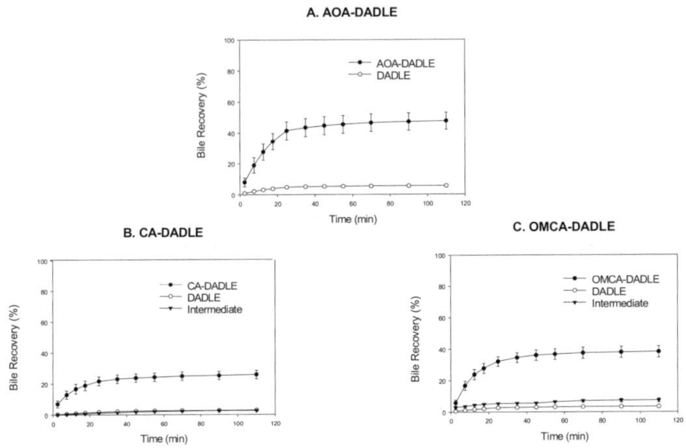

Figure 22. Time course for appearance of the DADLE prodrugs and their metabolites in bile after i.v. administration of the prodrugs to rats.

Table 6. Appearance of OMCA-DADLE and its metabolites in rat and guinea pig bile.

		Bile Recovery of Dose (%, mean ± SE)	
Species	**Compound**	**Prodrug**	**Linear Peptide**[b]
Rat[a]	OMCA-DADLE	38.1 ± 2.1	3.3 ± 0.4
Guinea Pig	OMCA-DADLE	3.3 ± 0.4	0.11 ± 0.01

a. Yang *et al.* [48]
b. DADLE

- **Conclusions about the DMPK properties of cyclic prodrugs of DADLE.**
  From the experimental data described above, one can arrive at the following conclusions about these cyclic prodrugs of DADLE: (i) they all have favorable physicochemical properties (e.g. hydrophobicity, low hydrogen-bonding potential, no charge) for cell permeation; (ii) they all exhibit good "intrinsic" cell permeation characteristics; (iii) they all are substrates for esterases that can catalyze their bioconversion to the DADLE; (iv) in some species the cyclic prodrugs are bioconverted more rapidly in brain than in blood; (v) in some animal species (e.g. rat) the cyclic prodrugs are substrates for efflux transporters (e.g. Pgp) that limit their permeation across the intestinal mucosa and the BBB; (vi) in some animal species (e.g. rat) the cyclic prodrugs are cleared rapidly from blood by the liver and appear in bile; and (vii) they all are substrates for cytochrome P450 enzymes, which also contribute to their high clearance.

Overall Conclusions

In the past 10 years, it has become very apparent to pharmaceutical companies that the high attrition rates for drug candidates in preclinical and clinical development were due in large part to drug candidates that possess poor "drug-like" properties. Today, companies have put into place "developability" criteria that restrict the advancement of molecules into preclinical and clinical development if they exhibit poor "drug-like" properties. Increased emphasis has been placed on optimizing the "drug-like" properties of molecules during the "hit-to-lead" and "lead" optimization stages in drug discovery.

This new drug discovery paradigm requires a new level of cooperation between scientists in discovery research and preclinical development units. To successfully implement this drug discovery paradigm, new scientific technologies need to be developed, validated, and integrated early in drug discovery. However, another very important ingredient for success is the scientist who works at the interface between discovery research and preclinical development. These individuals need to have scientific breadth, good communication skills, and respect for colleagues in other scientific disciplines. Discovery scientists need to develop the attitude that their job is to deliver drug candidates that meet the company's "developability" criteria so that these molecules have a high probability of success in the rigorous world of preclinical and clinical development. Preclinical development scientists need to view discovery scientists as their customers and thus help to educate them about the importance of meeting these "developability" criteria and help them design drug candidates that have optimal "drug-like" properties. If this new drug discovery paradigm is successfully implemented, the pharmaceutical industry is likely to see less attrition of its drug candidates in preclinical and clinical development in the future.

References

[1] J.A. DiMasi. New drug innovation and pharmaceutical industry structure: trends in the output of pharmaceutical firms. *Drug Inf. J.* **34** (2000) 1169-1194.

[2] S.A. Biller, L. Custer, K.E. Dickinson *et al.* The challenge of quality in drug optimization. In: R.T. Borchardt, E.H. Kerns, C.A. Lipinski *et al.* (Eds): *Pharmaceutical Profiling in Drug Discovery for Lead Selection.* AAPS Press, Arlington, VA, USA (2004) 413-430.

[3] T. Kennedy. Managing the drug discovery/development interface. *Drug Discov. Today* **2** (1997) 436-444.

[4] J.A. DiMasi. New drug development in the United States from 1963 to 1999. *Clin. Pharmacol. Ther.* **69** (2001) 286-296.

[5] J.A. DiMasi. Risks in new drug development: approval success rates for investigational drugs. *Clin. Pharmacol. Ther.* **69** (2001) 297-307.

[6] R.T. Borchardt. Scientific, educational, and communication issues associated with integrating and applying drug-like properties in drug discovery. In: R.T. Borchardt, E.H. Kerns, C.A. Lipinski *et al.* (Eds): *Pharmaceutical Profiling in Drug Discovery for Lead Selection.* AAPS Press, Arlington, VA, USA (2004) 451-466.

[7] R. Hirschmann. Introduction. In: R.T. Borchardt, R.M. Freidinger, T.K. Sawyer *et al.* (Eds): *Integration of Pharmaceutical Discovery and Development: Case Histories.* Plenum Press, New York, NY, USA (1998) 1-5.

[8] R.T. Borchardt, E.H. Kerns, C.A. Lipinski *et al.* (Eds). *Pharmaceutical Profiling in Drug Discovery for Lead Selection.* AAPS Press, Arlington, VA, USA (2004).

[9] R.T. Borchardt, E.H. Kerns, M.J. Hageman *et al.* (Eds). *Optimizing the "Drug-Like" Properties of Leads in Drug Discovery.* AAPS Press, Arlington, VA, USA (2006).

[10] I.J. Hidalgo, T.J. Raub, R.T. Borchardt. Characterization of the human colon carcinoma cell line (Caco-2) as a model system for intestinal epithelial permeability. *Gastroenterology* **96** (1989) 736-749.

[11] C.A. Lipinski, F. Lombardo, B.W. Dominy *et al.* Experimental and computational approaches to estimate solubility and permeability in drug discovery and development settings. *Adv. Drug Deliv. Rev.* **46** (2001) 3-26.

[12] C.A. Lipinski. Drug-like properties and the causes of poor solubility and poor permeability. *J. Pharmacol. Toxicol. Methods* **44** (2000) 235-249.

[13] K.R. Horspool and C.A. Lipinski. Enabling strategies: advancing new drug delivery concepts to gain

the lead. *Drug Deliv. Tech.* **3** (2003) 34-46.

[14] M. Kansy, H. Fischer, S. Bendels *et al.* Physicochemical methods for estimating permeability and related properties. In: R.T. Borchardt, E.H. Kerns, C.A. Lipinski *et al.* (Eds): *Pharmaceutical Profiling in Drug Discovery for Lead Selection.* AAPS Press, Arlington, VA, USA (2004) 197-216.

[15] I.J. Hidalgo, K. Weinstein, P. Kardos *et al.* Cultured epithelial cell assays used to estimate intestinal absorption potential. In: R.T. Borchardt, E.H. Kerns, C.A. Lipinski *et al.* (Eds): *Pharmaceutical Profiling in Drug Discovery for Lead Selection.* AAPS Press, Arlington, VA, USA (2004) 217-234.

[16] J.W. Polli and C. Serabjit-Singh. *In vitro* cell-based assays for estimating the effects of efflux transporters on cell permeation. In: R.T. Borchardt, E.H. Kerns, C.A. Lipinski *et al.* (Eds): *Pharmaceutical Profiling in Drug Discovery for Lead Selection.* AAPS Press, Arlington, VA, USA (2004) 235-256.

[17] M.J. Zamek-Gliszczynski and K.L.R. Brouwer. *In vitro* models for estimating hepatobiliary clearance. In: R.T. Borchardt, E.H. Kerns, C.A. Lipinski *et al.* (Eds): *Pharmaceutical Profiling in Drug Discovery for Lead Selection.* AAPS Press, Arlington, VA, USA (2004) 259-292.

[18] T.J. Raub, B.S. Lutzke, P.K. Andrus *et al.* Early preclinical evaluation of brain exposure in support of hit identification and lead optimization. In: R.T. Borchardt, E.H. Kerns, M.J. Hageman *et al.* (Eds): *Optimizing the "Drug-Like" Properties of Leads in Drug Discovery.* AAPS Press, Arlington, VA, USA (2006) 355-410.

[19] J.H. Lin. Challenges in drug discovery: lead optimization and prediction of human pharmacokinetics. In: R.T. Borchardt, E.H. Kerns, C.A. Lipinski *et al.* (Eds): *Pharmaceutical Profiling in Drug Discovery for Lead Selection.* AAPS Press, Arlington, VA, USA (2004) 293-326.

[20] R.E. White. A comprehensive strategy for ADME screening in drug discovery. In: R.T. Borchardt, E.H. Kerns, C.A. Lipinski *et al.* (Eds): *Pharmaceutical Profiling in Drug Discovery for Lead Selection.* AAPS Press, Arlington, VA, USA (2004) 431-450.

[21] D.R. Thakker. Strategic use of preclinical pharmacokinetic studies and in vitro models in optimizing ADME properties of lead compounds. In: R.T. Borchardt, E.H. Kerns, M.J. Hageman *et al.* (Eds): *Optimizing the "Drug-Like" Properties of Leads in Drug Discovery.* AAPS Press, Arlington, VA, USA (2006) 1-23.

[22] V.J. Stella, R.T. Borchardt, M.J. Hageman *et al.* (Eds). *Prodrugs: Challenges and Rewards: Part 1.* AAPS Press, Arlington, VA, USA (2007).

[23] V.J. Stella, R.T. Borchardt, M.J. Hageman *et al.* (Eds). *Prodrugs: Challenges and Rewards: Part 2.* AAPS Press, Arlington, VA, USA (2007).

[24] V.J. Stella. A case for prodrugs. In: V.J. Stella, R.T. Borchardt, M.J. Hageman *et al.* (Eds): *Prodrugs: Challenges and Rewards: Part 1.* AAPS Press, Arlington, VA, USA (2007) 3-33.

[25] P. Artursson and P. Matsson. Absorption prediction. In: R.T. Borchardt, E.H. Kerns, C.A. Lipinski *et al.* (Eds): *Pharmaceutical Profiling in Drug Discovery for Lead Selection.* AAPS Press, Arlington, VA, USA (2004) 3-26.

[26] G.M. Pauletti, F.W. Okumu, R.T. Borchardt. Effect of size and charge on the passive diffusion of peptides across Caco-2 cell monolayers via the paracellular pathway. *Pharm. Res.* **14** (1997) 164-168.

[27] G.T. Knipp, N.F. Ho, C.L. Barsuhn *et al.* Paracellular diffusion in Caco-2 cell monolayers: effect of perturbants on the transport of hydrophilic compounds that vary in charge and size. *J. Pharm. Sci.* **86** (1997) 1105-1110.

[28] A. Bak, T.J. Siahaan, O.S. Gudmundsson *et al.* Syntheses and evaluation of the physicochemical properties of esterase-sensitive cyclic prodrugs of opioid peptides using an (acyloxy)alkoxy linker. *J. Pept. Res.* **53** (1999) 393-402.

[29] A. Bak, O.S. Gudmundsson, G.J. Friis *et al.* Acyloxyalkoxy-based cyclic prodrugs of opioid peptides: evaluation of their chemical and enzymatic stability as well as their transport properties across Caco-2 cell monolayers. *Pharm. Res.* **16** (1999) 24-29.

[30] V.J. Stella. Prodrug approaches to enhancing the oral delivery of poorly permeable drugs. In: V.J. Stella, R.T. Borchardt, M.J. Hageman *et al.* (Eds): *Prodrugs: Challenges and Rewards: Part 1.* AAPS Press, Arlington, VA, USA (2007) 37-82.

[31] B. Wang, K. Nimkar, W. Wang *et al.* Synthesis and evaluation of the physicochemical properties of esterase-sensitive cyclic prodrugs of opioid peptides using coumarinic acid and phenylpropionic acid linkers. *J. Pept. Res.* **53** (1999) 370-382.

[32] H. Ouyang, D.G. Vander Velde, R.T. Borchardt *et al.* Synthesis and conformational analysis of a coumarinic acid-based cyclic prodrug of an opioid peptide with modified sensitivity to esterase-catalyzed bioconversion. *J. Pept. Res.* **59** (2002) 183-195.

[33] O.S. Gudmundsson, G.M. Pauletti, W. Wang *et al.* Coumarinic acid-based cyclic prodrugs of opioid peptides that exhibit metabolic stability to peptidases and excellent cellular permeability. *Pharm. Res.* **16** (1999) 7-15.

[34] H. Ouyang, F. Tang, T.J. Siahaan *et al.* A modified coumarinic acid-based cyclic prodrug of an opioid peptide: its enzymatic and chemical stability and cell permeation characteristics. *Pharm. Res.* **19** (2002) 794-801.

[35] O.S. Gudmundsson, D.G. Vander Velde, S.D. Jois *et al.* The effect of conformation of acyloxyalkoxy-based cyclic prodrugs of opioid peptides on their membrane permeation. *J. Pept. Res.* **53** (1999) 403-413.

[36] O.S. Gudmundsson, S.D. Jois, D.G. Vander Velde *et al.* The effect of conformation on the membrane permeation of coumarinic acid- and phenylpropionic acid-based cyclic prodrugs of opioid peptides. *J. Pept. Res.* **53** (1999) 383-392.

[37] J. Hochman, Q. Mei, M. Yamazaki *et al.* Role of mechanistic transport studies in lead optimization. In: R.T. Borchardt, E.H. Kerns, M.J. Hageman *et al.* (Eds): *Optimizing the "Drug-Like" Properties of Leads in Drug Discovery.* AAPS Press, Arlington, VA, USA (2006) 25-47.

[38] F. Tang and R.T. Borchardt. Characterization of the efflux transporter(s) responsible for restricting intestinal mucosa permeation of an acyloxyalkoxy-based cyclic prodrug of the opioid peptide DADLE. *Pharm. Res.* **19** (2002) 780-786.

[39] F. Tang and R.T. Borchardt. Characterization of the efflux transporter(s) responsible for restricting intestinal mucosa permeation of the coumarinic acid-based cyclic prodrug of the opioid peptide DADLE. *Pharm. Res.* **19** (2002) 787-793.

[40] L.Z. Benet and C.L. Cummins. The drug efflux-metabolism alliance: biochemical aspects. *Adv. Drug Deliv. Rev.* **50** (2001) S3-S11.

[41] J.H. Lin and M. Yamazaki. Clinical relevance of P-glycoprotein in drug therapy. *Drug Metab. Rev.* **35** (2003) 417-454.

[42] J.H. Lin and M. Yamazaki. Role of P-glycoprotein in pharmacokinetics: clinical implications. *Clin. Pharmacokinet.* **42** (2003) 59-98.

[43] H. Glaeser and R.B. Kim. The relevance of transporters in determining drug disposition. In: R.T. Borchardt, E.H. Kerns, M.J. Hageman *et al.* (Eds): *Optimizing the "Drug-Like" Properties of Leads in Drug Discovery.* AAPS Press, Arlington, VA, USA (2006) 423-459.

[44] R. Griffiths, A. Lewis and P. Jeffray. Models of drug absorption *in situ* and in conscious animals. In: R.T. Borchardt, P.L. Smith, G. Wilson (Eds): *Models for Assessing Drug Absorption and Metabolism.* Plenum Press, New York, NY, USA (1996) 67-84.

[45] Q.R. Smith. Brain perfusion systems for the studies of drug uptake and metabolism in the central nervous system. In: R.T. Borchardt, P.L. Smith, G. Wilson (Eds): *Models for Assessing Drug Absorption and Metabolism.* Plenum Press, New York, NY, USA (1996) 285-307.

[46] W. Chen, J.Z. Yang, R. Andersen *et al.* Evaluation of the permeation characteristics of a model opioid peptide, H-Tyr-D-Ala-Gly-Phe-D-Leu-OH (DADLE) and its cyclic prodrugs across the blood brain barrier model. *J. Pharmacol. Exp. Ther.* **303** (2002) 849-857.

[47] B.M. Liederer, P.K. Phan, H. Ouyang *et al.* Significant differences in the disposition of cyclic prodrugs of opioid peptides in rats and guinea pigs following IV administration. *J. Pharm. Sci.* **94** (2005) 2676-2687.

[48] J.Z. Yang, W. Chen, R.T. Borchardt. *In vitro* stability and *in vivo* pharmacokinetic studies of a model opioid peptide, H-Tyr-D-Ala-Gly-Phe-D-Leu-OH (DADLE), and its cyclic prodrugs. *J. Pharmacol. Exp. Ther.* **303** (2002) 840-848.

Towards Drugs of the Future
C.G. Kruse and H. Timmerman (Eds.)
IOS Press, 2008

doi:10.3233/978-1-58603-949-3-75

Hypothesis Generation for Scientific Discovery. Examples from the Use of KEM®, a Rule-Based Method for Multi-Objective Analysis and Optimization

Nathalie Jullian, Nathalie Jourdan and Mohammad Afshar
Ariana Pharma, BioTop, Institut Pasteur, 28 rue Dr Roux, 75724 Paris CEDEX 15, France

Abstract. Hypothesis generation is an essential step in scientific discovery. It involves an analysis of existing evidence, the generation of a "theory" (or model) which leads to a hypothesis (about the unknown) tested by new experiments. The design of new drugs follows a similar scheme starting with an effective mining of a huge amount of collected experimental *in vitro* and *in vivo* data. These data often come from many different areas such as chemistry, biology, pharmacology, toxicology etc. and in various formats. Extracting the critical information is a challenging task that is performed by medicinal chemists and other scientists. Our goal is to assist the hypothesis generation and decision process by mimicking as much as possible the human reasoning. To do so, we have developed a decision-support system KEM (Knowledge Extraction and Management), based on the Galois lattices theory. KEM assists the scientist for efficiently generating and managing consistent hypothesis from experimental data. KEM® is a machine learning software that can guide multi-objective optimisation. We present here a few examples of data analysis by using KEM.

Keywords. Data mining, QSAR, drug discovery, machine learning, multi-objective optimization

Introduction

Hypothesis generation is an essential step in scientific discovery. It involves an analysis of existing evidence, the generation of a "theory" (or model) which leads to hypothesis (about the unknown) tested by new experiments. This is an iterative cycle starting with mining of existing data and gradually validating or refuting hypotheses as they are generated. In the pharmaceutical industry, this process is clearly seen in the steps converting a "hit" molecule into a "lead" and further down the path into a "candidate" of interest for a therapeutic use. These molecules have to fulfil multiple and often competitive requirements related to better efficacy and safety [1-4]. The experimental evidence collected for each compound during this process increases dramatically, emphasising the need for smart and efficient data mining tools [5-8].

The data mining methodology implemented in KEM starts by building an ontology from complex data. The process derives knowledge in an exhaustive manner, learning from the entire dataset, including outliers and detecting non-statistically significant relations. The ontology, described as a set of rules, is used for extracting SAR information for multi-objective optimization and for end-point prediction.

The main advantages of this novel approach are:

- The dynamic handling of a large number of objectives,
- The capacity to work with a small number of examples that could be incomplete (i.e. with missing parameters),
- The exhaustive mining of the constraint set and the solution set,
- The ability for the user to examine the constraints and manipulate them in order to dynamically modify the knowledge base, while maintaining consistency.

Decisions/predictions can be explained and the analysis of contradictions allows the user to further understand the data, leading to an interactive exploration of the solution space. The origins of contradictions can be precisely identified and their identification is put at the heart of the scientific discovery process.

Multi-Objective Analysis

The design of effective and safe drugs involves optimizing multiple objectives at the same time. It is a common problem that the individual criteria can be conflicting or competing. A typical dataset contains multiple parameters such as:

- Multiple kinase assays: Activity1, Activity2, … etc.
- Volume of distribution values (L/kg)
- Functional Activities
- mwt (Molecular Weights)
- LogD pH 7.4 values
- T ½ (h) values
- F% (Fraction Absorbed) values
- multiple CYP panel: P450_1, P450_2, P450_3, … etc. data
- presence or absence of particular chemical fragments or atoms
- …

Since KEM uses class variables, continuous data fields are binned. For example, the activity will be described by 3 bins: Activity1_L (≤10 μM), Activity1_M (>10 μM, ≤1 nM), Activity1_H (>1 nM). The internal output of the system is a lattice from which rules are extracted. Each molecule can be characterized by a set of parameters displayed on a grid as shown in Figure 1.

i	u	Term	+	-	P	i	u	Term	+	-	P	i	u	Term	+	-	P	i	u	Term	+	-	P
0		Activity1_L		−	−	15		P450 1_L		−	−	30		P450 6_L		−	−	45		NH	✓		✓
1		Activity1_M		−	−	16		P450 1_M		−	−	31		P450 6_M		−	−	46		logD7.4_L		−	−
2		Activity1_H	✓		✓	17		P450 1_H	✓		✓	32		P450 6_H	✓		✓	47		logD7.4_M		−	−
3		T1/2 (h)_L		−	−	18		P450 2_L		−	−	33		Bromine atoms	✓		✓	48		logD7.4_H	✓		✓
4		T1/2 (h)_M		−	−	19		P450 2_M		−	−	34		Fluorine atoms		−	−	49		mwt_L		−	−
5		T1/2 (h)_H	✓		✓	20		P450 2_H	✓		✓	35		N in aromatic bonds with C	✓		✓	50		mwt_M	✓		✓
6		Vdss (L/kg)_L	✓		✓	21		P450 3_L		−	−	36		N non-ring bonded to a ring	✓		✓	51		mwt_H		−	−
7		Vdss (L/kg)_M		−	−	22		P450 3_M		−	−	37		heteroatoms in 5 ring	✓		✓	52		logP(o/w)_L		−	−
8		Vdss (L/kg)_H		−	−	23		P450 3_H	✓		✓	38		halogens bonded to any ring	✓		✓	53		logP(o/w)_M	✓		✓
9		F%_L		−	−	24		P450 4_L		−	−	39		N attached to CH2	✓		✓	54		logP(o/w)_H		−	−
10		F%_M		−	−	25		P450 4_M		−	−	40		chlorine atoms		−	−	55		pyrimidine		−	−
11		F%_H	✓		✓	26		P450 4_H	✓		✓	41		O attached to CH2		−	−	56		aminopyridine		−	−
12		Activ Funct_L		−	−	27		P450 5_L		−	−	42		N in rings	✓		✓	57		benzamidazole		−	−
13		Activ Funct_M		−	−	28		P450 5_M		−	−	43		het-het bonds		−	−	58		benzamide		−	−
14		Activ Funct_H	✓		✓	29		P450 5_H	✓		✓	44		non ring O connected to a ring		−	−	59		benzofurane	✓		✓

Figure 1. Grid showing the profile of a given compound X and its parameters in the ontology. Terms that have multiple "bins" are shown, for example Activity1 can be Low, Medium or High (Activity1_L, M, H). The "minus" sign shows that the value is not true; the green "checked" sign shows that the value is true. Values that are circled correspond to the experimental values. The "squared" values correspond to the desired values.

The grid can be used for propagating a specific profile aimed at reaching a combination of objectives, for example: Activity1_H, T1/2_H, F%_H and Activ_Funct_H. In this example, a set of rules have been derived from the lattice for 300 molecules. Propagating the "desired" multi-objective profiles leads to two "contradictions". The contradictions are of the type:

- If F%_H+ → benzofurane –
- If F%_H+ AND Bromine_atoms+ → Activ_Funct_L+

The first rule can be interpreted as: a benzofurane should not be present if we expect a high Fraction Absorbed (F%). The second rule is read as: the Functional Activity will be low when the molecule has a Br-atom and a high Fraction Absorbed. It identifies Br (in the context of a particular dataset) as the key "functional group" that hinders reaching both Fraction Absorbed High and Functional Activity High.

Decision Support in Lead Optimization

An experienced medicinal chemist analysing an SAR table automatically starts by generating hypothesis which are based on consistency. Often the hypothesis are based on observations which are not statistically significant, however they are not contradicted. One of our goals is to automate this process and to perform it in a systematic way. We present here an example of application for optimization of a series of di-substituted piperidine sigma ligands (Figure 2). We will consider a set of 102 molecules with activity data against the D2 and sigma receptors [9].

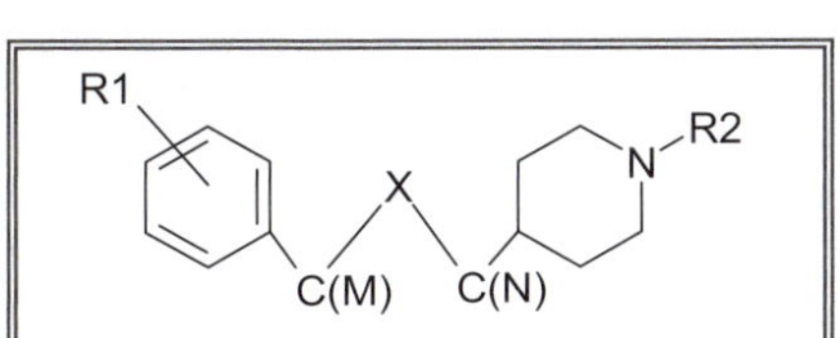

Figure 2. Chemical scaffold for the substituted piperidines. The molecules are described as lists of R1, R2, X, M and N.

There are two ways for exploiting the rules lattice with enumerative variables such as R-groups. Here we report the "Medium to Good" tool which proposes structural modifications to molecules with medium D2 activity, with the aim of turning them into molecules with high D2 activity. The ontology rules are applied for extracting a list of specific suggestions for chemical modifications. The figure below (Figure 3) illustrates a few chemical changes that *should* improve the activity against the dopamine D2 receptor. For example, line 1 suggests for molecule "10d" the introduction of a carbonyl group at the X position. An experienced chemist would be able to spot most of these changes. KEM allows this analysis to be done instantly and exhaustively. In addition, it is possible to look at multiple changes (i.e. changing 2 or more R-groups at a time) which are more difficult to track "by the human eye".

[From]	[X]	[M]	[N]	[R1]	[R2]
10d		[1+]	[0+]	[4F+]	[CH24Ph+]
10a	[CHOH+]	[1+]	[0+]	[4F+]	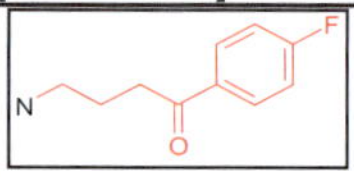
10d	[CHOH+]	[1+]	[0+]	[4F+]	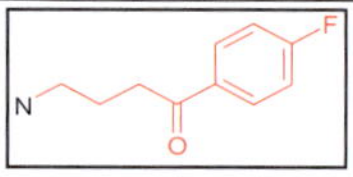
18ap	[O+]	[1+]	[2+]*	[4H3CO2C+]	[CH2Ph+]
18be	[O+]	[3+]	[2+]*	[H+]	[CH2Ph+]
6c	[CO+]	[0+]	[2+]*	[4MeS+]	[CH2Ph+]
18aw	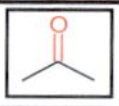	[1+]	[1+]	[4F+]	[CH2C6H4OCH2Php+]
18o	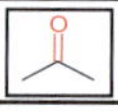	[0+]	[1+]	[4H3CCO+]	[CH2cC3H5+]

Figure 3. The column "From" points towards the identifier of the original molecule in the dataset. The suggested modifications are highlighted in red.

Analysis of Local QSAR Models

Although QSAR tools have been used for many years for supporting drug discovery efforts, they have been faced with two major challenges: (1) how to convert a QSAR equation into a suggestion for a new molecule (see the previous section) and (2) how to handle datasets where it is not possible to have a global model (i.e. high rates of false positive and/or negative). In short, how do we know whether we can predict the activity of a new molecule?

We show here an application of KEM as a predictive tool when global QSAR models are not possible. The fundamental steps in QSAR modelling are the definition of the application domain in the space of the chemical descriptors, the robustness of the model and its predictive power.

KEM is used to:

- Independently evaluate the predictability for active and inactive compounds.
- Immediate evaluation of the most pertinent descriptors at the learning step.
- Predict only "predictable" compounds from the testing sets.

In this example, a set of 397 DHFR (Di-Hydro-Folate Reductase) inhibitors is randomly and equally divided into a training set and a testing set [10]. It is important to note that there are only 25% of active compounds in the entire dataset.

When we examine the outcome of the prediction, there are 23% of false negatives with a numerical model such as Random Forest, a number close to the 25% expected if all compounds were predicted inactive [11]. There is indeed an over-representation of inactive compounds in this dataset, leading to an over-prediction for this class of activity. By using KEM specific rules, we propose local models for the individual prediction of activity and inactivity. The model predicts only 50% of the testing set, but the number of false negative decreases to 7% (Figure 4).

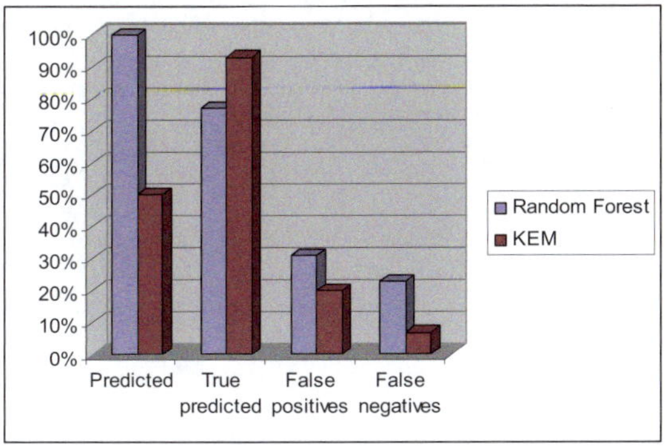

Figure 4. Evaluation of the quality of the prediction by using Random Forest (blue) and KEM (purple).

Before the model building, KEM allows an immediate evaluation of the pertinence of the descriptors at the learning step and indicates how many compounds from the training set are really used to build the model. The predictions are made only for a subset of compounds for which the system has "knowledge" hence improving the prediction quality (smaller number of false negatives and false positives).

Conclusion

The methodology that has been presented here mines knowledge from the data extracting rules in an exhaustive manner, learning from the entire dataset, including outliers and detecting non-statistically significant relations. The ontology, described as a set of rules, is used for extracting SAR information, for multi-objective optimization and for end-point prediction. KEM logic based system enables the automatic development of multiple local models, hence improving the reliability of the predictions.

This tool can be used at the heart of the hypothesis generation step in the scientific discovery process. Generating rules in an exhaustive way allows one to map the "known known" space. The hypotheses allow us to project into the "known unknowns". And as for the "unknown unknowns", the previous steps allow us to recognise them when we see them, i.e. be surprised!

References

[1] J.A. Kramer, J.E. Sagartz, D.L. Morris. The application of discovery toxicology and pathology towards the design of safer pharmaceutical lead candidates. *Nat. Rev. Drug Discov.* **6** (2007) 636-649.

[2] H. van de Waterbeemd and E. Gifford. ADMET in silico modelling: towards prediction paradise? *Nat. Rev. Drug Discov.* **2** (2003) 192-204.

[3] S. Ekins, J. Shimada, C. Chang. Application of data mining approaches to drug delivery. *Adv. Drug Deliv. Rev.* **58** (2006) 1409-1430.

[4] M. Afshar, A. Lanoue, J. Sallantin. Multi-objective/multi-criteria optimization and decision support in drug discovery. *Compr. Med. Chem. II* **4** (2007) 767-774.

[5] D.B. Searls. Data integration: challenges for drug discovery. *Nat. Rev. Drug Discov.* **4** (2005) 45-58.

[6] W. Loging, L. Harland, B. Williams-Jones. High-throughput electronic biology: mining information for drug discovery. *Nat. Rev. Drug Discov.* **6** (2007) 220-230. Erratum in: *Nat. Rev. Drug Discov.* **6** (2007) 442.

[7] S. Ekins, B. Boulanger, P.W. Swaan *et al.* Towards a new age of virtual ADME/TOX and multidimensional drug discovery. *J. Comput. Aided Mol. Des.* **16** (2002) 381-401.

[8] T.K. Sawyer. Smart drug discovery leveraging innovative technologies and predictive knowledge. *Nat. Chem. Biol.* **2** (2006) 646-648.

[9] P.J. Gilligan, G.A. Cain, T.E. Christos *et al.* Novel piperidine sigma receptor ligands as potential antipsychotic drugs. *J. Med. Chem.* **35** (1992) 4344-4361.

[10] J.J. Sutherland, L.A. O'Brien, D.F. Weaver. Spline-fitting with a genetic algorithm: a method for developing classification structure-activity relationships. *J. Chem. Inf. Comput. Sci.* **43** (2003) 1906-1915.

[11] V. Svetnik, A. Liaw, C. Tong *et al.* Random forest: a classification and regression tool for compound classification and QSAR modeling. *J. Chem. Inf. Comput. Sci.* **43** (2003) 1947-58.

Towards Drugs of the Future
C.G. Kruse and H. Timmerman (Eds.)
IOS Press, 2008

doi:10.3233/978-1-58603-949-3-81

Does Size Matter? Facts and Thoughts on Recombinant Drugs

Theodor Dingermann
Institute for Pharmaceutical Biology, Goethe-University, Frankfurt/Main and Center for Drug Research, Development, and Safety (ZAFES), Biocenter, Max-von-Laue-Strasse 9, D-60438 Frankfurt, Germany

Abstract. Since the approval of insulin in 1982 as the first recombinant protein, an impressive development has taken place. Meanwhile, more than 120 of recombinant drug substances have been approved and become available as extremely valuable therapeutic options. Based on a somewhat artificial but useful system one can classify recombinant drugs as first generation (changes due to technological concessions), second generation (authentic biomolecules), third generation (deliberately introduced modifications in order to improve mainly pharmacokinetic properties) and fourth generation drugs (newly invented, artificial proteins). Authenticity - exact copying of the most common human form - is not anymore a value per se, as challenges primarily related to the pharmacokinetics of artificial recombinant drugs can be overcome by diverging from the original. On the other hand, relatively minor changes in manufacturing or packaging may impact safety of therapeutic proteins.

Keywords. Recombinant drugs, production platforms

1. The Beginning

In 1973, Stanley Cohen and Herbert Boyer were the first to perform a type of experiment, which initiated, what was later coined recombinant DNA technology [1]. They lifted human DNA and copied it into a bacterial plasmid, which they successfully introduced into a bacterium, resulting in this genetically engineered bacterium producing a foreign protein. This was possible because the building blocks of DNA, and the rules for encoding biological information by DNA (genetic codes), are identical among different species. Only those regions which control the transcription process of the DNA-encoded information into mRNA have to be adopted to the target organism, since expression control of genetic information is highly specific not only per species but frequently even per specific cell-type.

2. From a Pharmaceutical Perspective

From a pharmaceutical perspective, this successful recombinant DNA experiment opened the possibility to generate human biomolecules in any other organism. Herewith, an alternative to isolation of therapeutic proteins from biological materials (tissues, organs), in terms of production capacity and/or safety, was now provided.

The first therapeutic protein produced by recombinant DNA technology was human insulin, for which the first marketing authorization was obtained in 1982, just nine years after Cohen's and Boyer's pioneering experiment [2].

Human proteins produced through genetic engineering rather than isolated from tissue samples have become an important category within therapeutic medicines. Progress in production technology, as well as in pharmacological understanding, will allow for the continued development of human proteins as an important therapeutic option in a variety of human diseases.

3. From a Clinical Perspective

From a clinical perspective, therapeutic proteins provide essential therapies in various disease including diabetes (insulin), end-stage renal disease (erythropoietin), viral hepatitis (interferon), cancer-treatment related neutropenia (G-CSF), clotting disorders (factor VII, VIII, IX), inborn errors of metabolism (lysosomal enzymes), chronic inflammation (TNF-alpha blockers), multiple sclerosis (beta-interferons), and many others.

4. From a Biochemical Viewpoint

From a biochemical viewpoint, there is an interesting diversity within these proteins in their primary protein structure (chain of amino acids) compared to the naturally occurring human form(s) and their post-translational modifications (including the natural process of glycosylation, the addition of sugar groups to specific amino acids, and the artificial process of pegylation, the addition of polyethylene glycol [PEG] groups to amino acids). Initially, biochemical differences between the therapeutic proteins and their natural counterparts often reflected either technical or biological limitations. Such deviations, however, do not automatically have an actual impact on their therapeutic value. Thus, a goal of a therapeutic protein being as identical as possible to the human counterpart is a common misconception. Increasingly, differences represent deliberate modifications meant to mainly improve pharmacokinetic properties, or, in rare cases, to alter pharmacodynamic characteristics. Pegvisomant (Somavert™), for example, is a human growth factor variant with nine amino acids substituted causing a switch from an agonist to an antagonist, which found its indication for the treatment of acromegaly. Thus, the therapeutic potential and value is not determined by the resemblance between the therapeutic protein and the naturally occurring human protein.

5. From a Safety Viewpoint

From one safety point of view, it would theoretically be best to choose an expression platform that is phylogenetically most distant from humans since human pathogens cannot contaminate such systems. If glycosylation is required, then normally mammalian expression platforms are favorable, such as CHO- or BHK-cells. Any expression system has to prove an acceptable safety profile. This is also one reason, why approved drug substances originate from a fairly limited number of expression systems.

6. The Paradigm Change in Drug Definition

In a monograph entitled "RECOMBINANT DNA TECHNOLOGY, PRODUCTS OF" (Producta ab ADN recombinante) the European Pharmacopeia defines recombinant products as follows:

> "Products of rDNA technology are produced by genetic modification in which DNA coding for the required product is introduced, usually by means of a plasmid or a viral vector, into a suitable micro-organism or cell line, in which that DNA is expressed and translated into protein. The desired product is then recovered by extraction and purification. ..."

According to this definition recombinant products are always proteins. But more importantly (according to the second sentence) recombinant drugs are defined based on their concrete production process. The new paradigm reads: "The product is the process" (Figure 1).

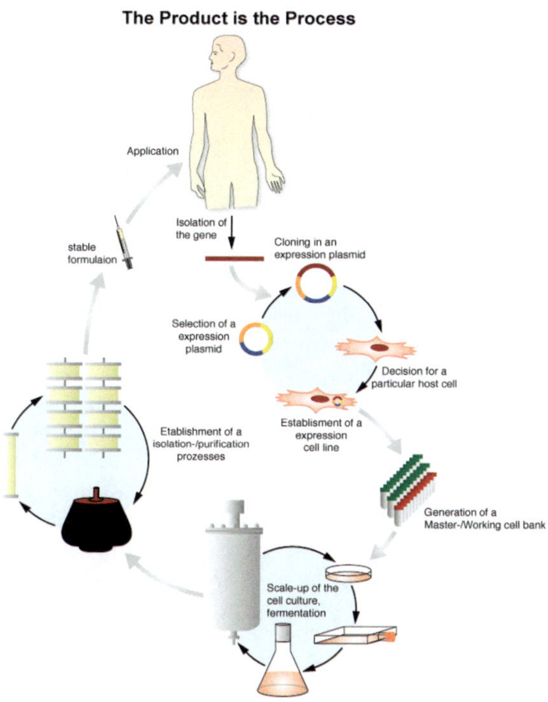

Figure 1. "The product is the process".

While low molecular weight drug substances are sufficiently defined and characterized by chemical and physical properties, the definition of biotechnological drug substances includes in addition any relevant process parameter. These parameters have to be defined in advanced, and based in these definitions the process is specified. Any deviation from these parameters causes an "out of specification (OAS)" result, which objects against approval regardless whether the quality of the drug substance is affected. This guarantees a remarkable reproducibility of recombinant drug's quality, which is of ad most importance since proteins can undergo denaturation and activity loss just on the basis of three-dimensional structure changes without any changes in the primary structure.

7. Protein Production Platforms

The first expression platform established was *Escherichia coli* (*E. coli*), which came with a large body of knowledge on the genetics, simple cultivation requirements and a short generation time. This platform worked well for insulin and human growth hormone. But soon it was realized that *E. coli* had one serious shortcoming: it was unable to modify proteins by glycosylation. This was not a problem for producing insulin and human growth hormone since these proteins do not undergo glycosylation in their natural human form. Problems were, however, anticipated for many other recombinant proteins with therapeutic value, since most of the human extracellular proteins (and even some intracellular proteins) are naturally modified with sugar chains.

Glycosylation of proteins is a highly complex post-translational modification process taking place in the endoplasmatic reticulum and Golgi apparatus and involving more than a hundred different proteins (and genes). This glycosylation machinery is absent in *E. coli*, present but different from mammalian cells in the yeast *Saccharomyces cerevisiae*, but highly conserved among mammalian cells (e.g. human, Chinese hamster).

It came as a welcome surprise that even some therapeutic proteins like the interferons alfa, beta and gamma, interleukin 2, tumor necrosis factor alfa and others, which are glycosylated in their natural human forms, turned out to be effective medicines, even though they were not glycosylated after expression in *E. coli*.

However, other proteins failed as therapeutics when not glycosylated. The most prominent in this group was erythropoietin, which when originally produced in *E. coli* was fully active *in vitro* but not efficacious *in vivo* [3]. The cause was an insufficient pharmacokinetic profile due to the absence of glycosylation, and not a primary pharmacodynamic failure, as receptor binding itself is not dependent on sugar modifications. Whereas there are two forms of glycosylation of proteins, O- and N-glycosylation, only the latter is relevant for all human proteins that have been reproduced to date. N-linked glycosylation involves the addition of sugar groups (glycans) to the amino acid asparagine within the polypeptide chain.

Although the yeast *Saccharomyces cerevisiae* expresses N-glycosylation sites encoded by human genes, its glycosylation pattern is so different from that of humans [4] that concerns whether such an "artificial" glycosylation of recombinant products would induce a severe immune response in humans were raised. Nevertheless, *S. cerevisiae* is meanwhile an established host organism for a number of recombinant proteins (human insulin, glucagon, the hirudin analogues desirudin and lepirudin, the urate oxidase rasburicase, and the platelet-derived growth factor [PDGF] becaplermin, and Hepatitis B-antigens used for vaccination), where the yeast glycosylation is not an issue in clinical practice.

Use of three mammalian cell lines, derived from Chinese Hamster Ovary cells (CHO-cells), Baby Hamster Kidney cells (BHK-cells), or human fibrosarcoma cells, elegantly provided the necessary glycosylation in the production of many other therapeutically intended naturally glycosylated proteins.

The mammalian cell systems have an additional advantage: recombinant proteins are secreted into the media in a natural form. Proteins expressed in *E. coli*, on the other hand, mostly accumulate protein products within the cell as inclusion bodies in a highly denaturated form, from which they have to be renaturated during further manufacturing - generally with limited although sufficient yields.

For a long period of time, basically all recombinant proteins which were approved as drugs were produced in just four cell systems: *E. coli*, *S. cerevisiae*, CHO- and BHK-cells. Frequently, a particular therapeutic protein was available from two different expression systems. Insulin and human growth hormone, for example are expressed either in *E. coli* or in *S. cerevisiae*. Interferon beta is expressed in *E. coli* and in CHO-cells. In this latter case, the final products even differ structurally. While the CHO-produced interferon beta is amino acid-wise identical to the human original, the *E. coli* produced interferon beta is one N-terminal amino acid shorter, has a cysteine substituted by a serine, and lacks any glycosylation. So, interestingly, it is apparently not possible to anticipate superiority for the more "authentic" or "natural" version of the glycoprotein. A head-to-head clinical study [5], which compared the "artificial" interferon beta-1b and the "authentic" interferon beta-1a showed superiority in favor of the "artificial" interferon beta-1b. However, definite proof, that the one interferon molecule is indeed superior to the other cannot be deduced from this study since both drugs were given in different doses and at different dosing intervals. Interferon beta-1a induced less antibody formation and could therefore theoretically be regarded as better tolerated by the human body, however with no real clinical relevance shown in patient users.

8. Authentic *versus* Modified Recombinant Products

The authenticity of recombinant drugs is not regarded any more as an important advantage. In fact only a minority of all approved recombinant drugs correspond strictly to their "authentic" counterparts.

Besides the fact that early products, which were produced in *E. coli*, differed from the "original" due to technological concessions (e.g. lack of glycosylation, substitution of cysteine by serine, N-terminal amino acid addition or N-terminal amino acid loss), an important driver for intentionally introduced structural changes are the pharmacokinetic characteristics. Regardless of the choice of production platform, we are still far away from (and may probably never be able to) applying recombinant proteins in the exact same way as the body delivers them. Insulin, for example, is subtly secreted from the pancreatic beta-cells upon physiological stimuli, whereas it is administered by a syringe subcutaneously in a very high millimolar concentration. This concentration forces recombinant insulin into hexameric aggregates which are unable to bind to the receptor. Receptor binding can only occur after dilution by diffusion, with the consequence that patients are advised to anticipate an efficacy lack for 15 to 30 minutes post injection. This problem, which clearly interferes with patient compliance, was elegantly solved by slight structural modifications at the C-terminal region of the B-chain of the protein core. The resulting second-generation fast acting insulins were Lispro™, Aspart™ and Glulisin™. These fast acting insulins are functionally complemented by long acting insulin derivatives, again based on intelligent structural modifications. It is highly plausible that combinations of such modified insulins will prove beneficial in the long term, not as much based on molecular superiority, but based on superior patient compliance [6].

Another example of deliberate change from "authenticity" is Tenectaplase (a modified tissue specific plasminogen activator, tPA). In Tenectaplase, the plasminogen activator inhibitor-1 (PAI-1) binding site Lys-His-Arg-Arg was replaced by an Ala-Ala-Ala-Ala sequence. This resulted in a significantly longer biological half-life compared to native tPA due to greater resistance to biological inactivation [7].

Similarly, Reteplase (Rapilysin™) is an N-terminal truncated tPA-version; in addition, it is not glycosylated since it is produced in *E. coli*. This second-generation tPA was approved on the basis of favorable pharmacokinetic considerations, although the authentically produced human tPA Alteplase (Actilyse™) was already on the market.

We will see a great variety of third-generation biotechnology products as fragments from authentic human proteins in the future, many of them being antibody fragments or antibody-derived peptides.

9. Post-Expression Modification of Recombinant Proteins

Not only modifications in the polypeptide chain but also in the post-expression modification can provide the basis for second-generation biotechnology products. Modifying proteins with polyethylene glycol (PEG) chains, or with additional sugar chains, changes pharmacokinetic properties significantly, and frequently in favor of overall therapeutic efficacy. For example, PEGylated interferon alfa variants, with a lengthened plasma half-life, clearly proved to be therapeutically superior in treating chronic Hepatitis C and B infections compared to unmodified alfa-interferons.

Darbepoeitin with two additional glycosylation chains has a lower receptor affinity compared to unmodified erythopoietins but a significantly longer biological half-life. The overall therapeutic efficacy appears to be superior for the modified erythropoietins [8]. Darbopoeitin has a higher sialic acid content, providing more protection from hepatic uptake by the asialoglycoprotein receptors present on liver cells, as the result of the replacement of two amino acids by asparagine residues, thereby creating two additional sites for N-linked glycosylation. Darbopoeitin is produced in CHO-cells, and its excellent safety record associated with repeated administration in patients, provides support to the value of the CHO-platform to produce adequately glycosylated human proteins, including those with high(er) sialic acid contents.

In the case of glucocerebrosidase, therapeutic efficacy turned out to be dependent on a complex sugar modification. Glucocerebrosidase is a lysosomal enzyme, which is usually delivered through an intracellular delivery system starting at the Golgi apparatus and ending in the lysosomes. Gaucher disease is characterized by an inborn deficiency of this enzyme and it needs to be infused intravenously to reach macrophages (the main site of pathological lysosomal accumulation of glucocerebroside) in sufficient quantities. Surprisingly, this could not be accomplished with the natural form of glucocerebrosidase but, after it's sugar chains were cut back to expose mannose, such modified glucocerebrosidase actually was able to achieve therapeutic concentrations in target cells (monocytes and macrophages). Therefore, a complex biochemical modification of recombinant glycocerebrosidase sequentially using three specific glycosidases (neuraminidase, ß-galactosidase and N-acetyl-glucosaminidase) is required to obtain mannose-terminated glucocerebrosidase (Cerezyme™), which has proven to be a highly safe and successful therapy in patients suffering from Gaucher disease [9]. These interesting observations, which elegantly solved a long existing unmet medical need, cannot be extrapolated towards even higher efficacy by adding additional mannose groups. In a recent preclinical study, various expression and modification systems were used to produce different mannose-terminated glucocerebrosidases. The ones containing more mannose, and as such resembling proteins produced in yeast expression systems, were not more active than standard Cerezyme™ [10].

10. Immunological Considerations

A major concern for all recombinant drugs is immunogenicity. It can be assumed that all biotechnologically produced therapeutics may exhibit some form of immunogenicity during an unpredictable time scale. An antibody response can result in reduced efficacy of the therapeutic protein itself, or, very rarely, in inactivation of native endogenous proteins by neutralizing antibodies, or in anaphylactic reactions [11,12]. Well-known is the Eprex™ case [13]. Between 1998 and 2002, unusual high numbers of pure red cell aplasia (PRCA) were reported in patients with chronic kidney disease who received subcutaneously administered Eprex™. PRCA resulted from depletion of endogenous and therapeutically applied erythropoietin by neutralizing antibodies induced by the treatment. Most likely, the presence of leachates in the formulation, arising from an interaction between polysorbate 80 and the uncoated rubber stopper were responsible for the observed antigenicity.

This highlights the fact that safety of new therapeutic protein products is difficult to predict. The Eprex™ case taught us that the combination of the recombinant protein and formulation/packaging constituents can be more critical for safety than changes in the recombinant protein itself. On the other hand, the huge number of molecular changes compared to the natural originals in authorized biopharmaceuticals and their clinically demonstrated fairly safe therapeutic use has put immunogenicity with respect to safety concerns in perspective.

Overall safety, and in particular immunogenic safety, can only be assessed through clinical and post-marketing programs due to 1) the human immune system being more sensitive than the available physical tests or bioassays; 2) the limitations of current analytical methods; 3) the lack of standardized assays; and 4) a certain degree of idiosyncrasy of each individual's immune system and immune disposition. Rare immune-mediated reactions (e.g. 1 in 10,000 patient-years) will only become apparent through robust post-marketing surveillance.

Not yet adequately addressed is a potential reduction of efficacy of the therapeutic protein caused by neutralizing antibodies. If this is relevant, then this problem can be overcome by dosing consistently on the basis of antibody titers and not only on body weight. Authorities become more and more aware of this potential problem and may ask for studies in this direction.

11. Conclusions

Meanwhile the number of recombinant drugs exceeds 120. Based on the unique potential of biotechnology to copy any genetic information in any living organism or cell, the first molecules were indeed exact (or almost exact) copies of human proteins produced in insufficient quantities in particular patients (substitution drugs). However, after scanning the list, it is obvious that the landscape really changed. Recently developed biotechnologicals are exact copies of human proteins only in a very few cases. As one of the big surprises, these modified biotechnologicals are also fairly well-tolerated and mostly do not provoke immune response complications. This is good news, since modifications are frequently needed to improve pharmacokinetic properties of biotechnologicals.

References

[1] S.N. Cohen, A.C. Chang, H.W. Boyer *et al.* Construction of biologically functional bacterial plasmids in vitro. *Proc. Natl. Acad. Sci. U.S.A.* **70** (1973) 3240-3244.

[2] Anonymous. Human insulin receives FDA approval. *FDA Drug Bull.* **12** (1982) 18-19.

[3] H. Sasaki, B. Bothner, A. Dell *et al.* Carbohydrate structure of erythropoietin expressed in Chinese hamster ovary cells by a human erythropoietin cDNA. *J. Biol. Chem.* **262** (1987) 12059-12076.

[4] T.R. Gemmill and R.B. Trimble. Overview of N- and O-linked oligosaccharide structures found in various yeast species. *Biochim. Biophys. Acta* **1426** (1999) 227-237.

[5] L. Durelli, E. Verdun, P. Barbero *et al.* Every-other-day interferon beta-1b versus once-weekly interferon beta-1a for multiple sclerosis: results of a 2-year prospective randomized multicentre study (INCOMIN). *Lancet* **359** (2002) 1453-1460. Comment in: *Lancet* **360** (2002) 1428; author reply 1428-1429.

[6] D. Einhorn. Advances in diabetes for the millennium: insulin treatment and glucose monitoring. *MedGenMed.* **6 (3 Suppl)** (2004) 8.

[7] C.J. Dunn and K.L. Goa. Tenecteplase: a review of its pharmacology and therapeutic efficacy in patients with acute myocardial infarction. *Am. J. Cardiovasc. Drugs* **1** (2001) 51-66.

[8] A.M. Sinclair and S. Elliott. Glycoengineering: the effect of glycosylation on the properties of therapeutic proteins. *J. Pharm. Sci.* **94** (2005) 1626-1635.

[9] G.M. Pastores, N.J. Weinreb, H. Aerts *et al.* Therapeutic goals in the treatment of Gaucher disease. *Semin. Hematol.* **41** (2004) 4-14.

[10] S.M. van Patten, H. Hughes, M.R. Huff *et al.* Effect of mannose chain length on targeting of glucocerebrosidase for enzyme replacement therapy of Gaucher disease. *Glycobiology* **17** (2007) 467-478.

[11] M. Kessler, D. Goldsmith, H. Schellekens. Immunogenicity of biopharmaceuticals. *Nephrol. Dial. Transplant.* **21 Suppl 5** (2006) v9-v12.

[12] H. Schellekens. Factors influencing the immunogenicity of therapeutic proteins. *Nephrol. Dial. Transplant.* **20 Suppl 6** (2005) vi3-vi9.

[13] H. Schellekens and W. Jiskoot. Eprex-associated pure red cell aplasia and leachates. *Nat. Biotechnol.* **24** (2006) 613-624. Comment in: *Nat. Biotechnol.* **24** (2006) 1199-1200; author reply 1200.

Towards Drugs of the Future
C.G. Kruse and H. Timmerman (Eds.)
IOS Press, 2008

doi:10.3233/978-1-58603-949-3-89

Can Nanotechnology Improve the Oral Delivery of Therapeutic Proteins?

G.T. Robillard[a], K. Scholtmeijer[b], M. Haas[a], J. Smisterova[a]
[a] *Biomade Technology Foundation, Nijenborgh 4, 9747 AG Groningen, The Netherlands*
[b] *Department of Biology, University of Utrecht, Padualaan 8, 3584 CH Utrecht, The Netherlands*

Abstract. Achieving effective oral delivery of therapeutic proteins means circumventing physiological barriers which seem to have evolved specifically to prevent this process. There are two classes of barriers, the intestinal wall and enzymatic degradation. Permeation constraints at the intestinal wall are attributed to the large molecular size of the proteins which prevent the paracellular route of permeation while their high aqueous solubility/hydrophilicity greatly constrains transcellular diffusion. The GI tract, on the other hand, contains a variety of proteases designed to degrade large proteins all the way to di-and tri-peptides by the time they reach the basolateral side of the epithelial cell layer of the small intestine.

Many approaches have been and are being investigated to address the challenge of oral protein delivery, only a subset of which falls in the category of nanotechnology. This article treats a number of these nanotechnologies including micro-machined devices delivering nanoliter volumes, mesoporous silica nanoparticles, protein nano-vessels and receptor-mediated endocytotic delivery of protein nanoparticles.

Keywords. Micro-machined devices, mesoporous silica, receptor-mediated endocytosis, hydrophobin, vitamin B_{12} uptake

Introduction

A Google search, 18 December 2007, with the combined terms "nanotechnology" and "oral protein delivery" gives the surprisingly small number of 266 hits, about half of which are advertisements. Even more surprising is the result of the same search in the USPTO (U.S. Patent and Trademark Office) database [1]. Zero patent applications and zero issued patents. Given the explosive developments in nanotechnology (11,900,000 Google hits) and the generally accepted importance of oral protein delivery, it's obviously the link between the two topics which is limiting the number of hits. And so, it is legitimate to ask whether nanotechnology can do anything to improve the perspectives for oral protein delivery. Without trying to be exhaustive, this paper will give an overview of the issues confronting oral protein delivery and examples of the some of the routes being followed in looking for solutions, especially as they concern nanotechnological approaches.

Nanotechnologists use dimensions to distinguish their world from the molecular world, nano-objects being complexes with dimensions from one to several hundred nanometers. When it comes to proteins this is not a very useful classification since proteins are often composed of multiple subunits each of which can have a diameter of several nanometers while the resulting functional complex can have a cross-section of several tens of nanometers. Thus, in the following pages, all but the smallest proteins can be considered to be nanoparticles, but that, in itself, does not answer the question in the title line. The issue is whether there are approaches specific to this field of technology which can enhance the movement of an individual protein or aggregates of them from the gut to the blood stream.

Before addressing this issue, we need to define "nanotechnology". There are many definitions. Two, given by the British Royal Society and the U.S. National Nanotechnology Initiative, respectively, are:

> "Nanotechnologies are the design, characterisation, production and application of structures, devices and systems by controlling shape and size at nanometre scale" [2].
>
> "Nanotechnology is the ability to understand, control, and manipulate matter at the level of individual atoms and molecules, as well as at the "supramolecular" level involving clusters of molecules (in the range of about 0.1 to 100 nM), in order to create materials, devices, and systems with fundamentally new properties and functions because of their small structure" [3].

The common words appearing in these and in most other definitions of "nanotechnology" are "creation", "production", "application", "structures", "devices" and "systems" So, to answer the question in the title, we look at whether the application of structures, devices and systems in the nM range can improve the oral delivery of proteins. The challenge is not so much the technology as the complexity of the physiological environment and processes involved.

Factors Affecting Protein Uptake in the Gut

The gut is a complex environment, a mixture of saliva and food components, commensal gut flora, mucus, immunoglobulins and other components excreted the epithelial cells, which interact with the surface of proteins and influence their movement through the mucus gel layer and interaction with the epithelial cell membrane.

The gut is a dynamic environment. Particles moving through the gut must attach to epithelial cells before they are excreted but to do so they must penetrate a mucous gel layer which covers the epithelial cells. Its composition is a function of the health status of the individual and its thickness is a function of the location along the GI tract. In all cases, the size and density of the particles are important factors in penetrating the mucous and adhering to the epithelial cell layer. Ease of penetration increases with increasing density but decreasing size. This also applies to their ability to adhere to the epithelial cell surface in the face of sheer forces which accompany mucosal gel layer flow.

The gut is a harsh environment that challenges the physical and chemical stability of proteins. If they survive the stomach and arrive intact in the intestine, they are exposed to a plethora of proteolytic enzymes intended to break them down into single amino acids or di- and tripeptides for adsorption by the intestinal epithelium through the action of specific transport proteins. Protection against these environmental factors is the function of most encapsulation technologies.

Routes for Drugs Crossing the Intestinal Membrane

There are several routes for transfer of components from the gut across the intestinal epithelial cell layer to the blood stream. (*i*) *Carrier-mediated transport* is the most selective route. Only those components are translocated which interact specifically with the carrier binding site. Proteins are not transported by such carriers, only their break-down products such as di- and tripeptides and single amino acids. There are, however, some examples of transporters which can transport their natural substrates when that substrate is attached to another cargo. (*ii*) *Passive diffusion* can occur either across the cell, transcellular, or in between adjacent cells, paracellular. To cross the membrane by passive diffusion, the *transcellular route*, the lipophilicity must be similar to that of the membrane components themselves. Unfortunately, anything that lipophilic will be very poorly water-soluble, aggregate and be restricted in diffusion across the membrane due to the large size of the complex. This is certainly the case for proteins and most large peptides. Hydrophilic proteins and peptides, on the other hand, would have to follow the *paracellular route* except that this route is very narrow, 1-5 nM in breadth and is blocked by the tight junction protein complexes, intended to seal of the separation between cells in order to prevent bacteria and toxins from following this route into the blood stream. Unless the tightness of these junctions can be modulated by external factors, the access of hydrophilic proteins and peptides larger than approximately 3 kDa is blocked. (*iii*) The *endocytotic and exocytotic route* involves many membrane and cytoplasmic components and regulatory events. The epithelial cell membrane facing the lumen of the intestine engulfs components outside of the cell, folds inwards and pinches off to form internal vesicles containing these components. These travel to the basolateral side of the cell fuse with the membrane thereby expelling the components to be taken into the blood. While there are several classes of endocytotic mechanisms, the focus here will be on receptor-mediated endocytosis.

Approaches to Enhance Oral Protein Delivery

The conditions in the gut and the characteristics of the intestinal epithelial cell layer define the approaches which are being taken to enhance oral protein delivery. These involve two classes, those aimed at helping the proteins survive the harsh environment of the gut and those helping to move the proteins past the epithelial cell layer to the blood stream
Non-specific approaches include the use of:

- Protease inhibitors including Ca^{++} chelators and carrier devices to protect the proteins from degradation on their journey through the gut.
- Various approaches for enhancing permeation through or between the epithelial cells such as:

- Cell-penetrating peptides.
- Permeation enhancers (transcellular & paracellular):
  - Ionic and non-ionic surfactants;
  - Bile salts;
  - Fatty acids and derivatives.
- Chemical conjugation of hydrophobic molecules to proteins, for example:
  - Hexyl-insulin monoconjugate-2 (Nobex).
- Small molecule carriers such as Emisphere's Eligen® technology.

Specific approaches focus on enhancing permeation and include:

- Nanoparticle formulations.
- Opening tight junctions through the use of *Vibrio cholera* Zonula occludens toxin (ZOT) & ΔG (the toxin's 12 kDa C-terminal fragment).
- Receptor-mediated uptake using, for instance:
  - Vitamin B_{12} nanosphere conjugates;
  - *Pseudomonas Aeruginosa A* endotoxin;
  - Polymeric receptor.

Clearly, many of the above approaches are not specific to nanotechnology. They are standard chemical approaches to solving the stability and penetration problem. However, delivery devices, nanoparticle formulations and systems designed for receptor-mediated delivery do qualify as nanotechnological approaches which offer real potential for improving oral protein delivery. Examples of each will be treated in the following sections.

Delivery Devices to Protect Against Degradation

Devices for the protection of proteins during oral delivery have been constructed in two ways, by using micro-fabrication procedures perfected in the electronics industry, and by self-assembly starting from molecular components. These two approaches are known in the nanotech world as the "top-down" and "bottom-up" approaches, respectively.

Desai and colleagues [4-6] have pioneered the micro-fabrication approach. Using standard photolithographic and etching procedures they have fabricated sheets of micron-sized vessels from substrates such as silicon dioxide and later separated them into individual vessels with pico-liter volumes (Figure 1, left and middle, respectively). In contrast to other drug delivery particles which are spherical, their devices are flat, thereby presenting a much larger surface for more efficient transfer of its protein content when attached to the epithelial cell layer (Figure 1, right). To enhance the residence time of the micro-devices on the intestinal epithelium, they were decorated with biotinylated tomato lecitin through an avidin coupling to amine-modified surfaces of the devices [6]. Tomato lecitin is known to target Caco-2 cells. Monolayers of these cells retained the modified devices during repeated washings and maintained their integrity for periods of exposure to the devices of up to 48 hours.

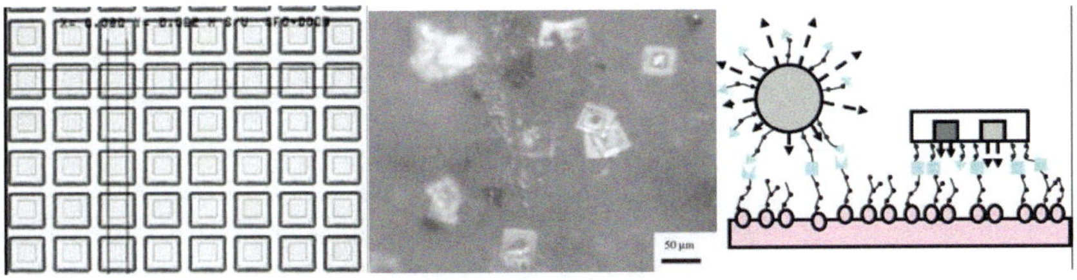

Figure 1. Reproduced from references [4] and [6], with permission.

Using the example of erythropoietin (EPO) therapy for treatment of anaemia, requiring i.v. administration three times per week, they calculated that, at an EPO concentration of 50 μg/ml and 100,000 U per mg, a 6.5 pl reservoir in the micro-device would hold 0.03 U of EPO. Allowing for only 33% oral bioavailability relative to i.v. administered EPO, an adult dose would require 300,000 micro-devices, corresponding to 100 mg of material, an amount easily contained in a capsule [4]. With the technology available such devices can be mass produced and should be able to serve the dual function of protecting therapeutic proteins from degradation in the gut as well as preferred presentation and release at the intestinal epithelium.

Mesoporous silica nanoparticles offer a bottom-up approach with similar potential for delivering and protecting orally administered proteins from degradation. The basic technology for forming mesoporous structures was developed by Mobil Research and Development Corporation in the early '90s [7,8]. Quarternary ammonium surfactants form hexagonal liquid crystal phases which serve as the template for the formation of hexagonal arrays of uniform channels in the presence tetramethylammonium silicate.

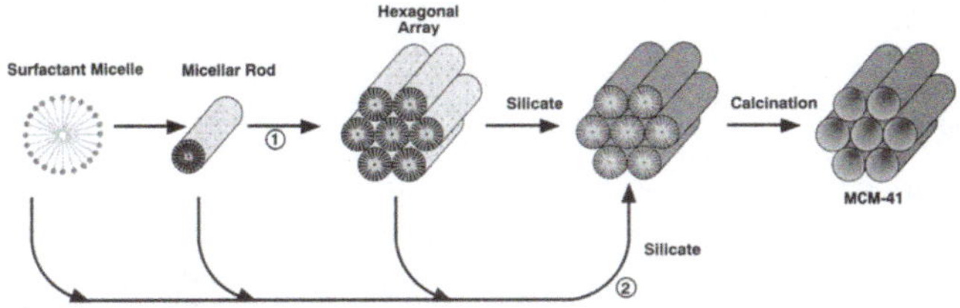

Figure 2. Possible mechanistic pathway for the formation of hexagonal mesoporous silica (reproduced from reference [7], with permission).

This family of hexagonal mesoporous molecular sieves, labelled MCM-41 (Mobile Composition of Matter-41), can be constructed with pore diameter which increase with increasing alkyl chain length from 1.8 nM (n = 8) to 3.7 nM (n = 16). Further increases up to 10 nM are achieved by addition of auxiliary organics such as 1,3,5-trimethyl benzene which expands the micelle size. The suggested mechanism for the formation of these structures is shown in Figure 2. After the silicate polymerizes around the micellar rods, heat is used to burn off the surfactants and auxiliary organics, leaving porous silicate structures. Lin and colleagues prepared MCM-41 type structures in the nanoparticle form and showed that they could be used for the controlled release of antibiotics [9] and subsequently, proteins [10], thereby opening the perspective to use such systems for transmembrane transport of proteins and eventually oral protein delivery.

Figure 3. Electron micrographs of large pore mesoporous silica nanoparticles (reproduced from reference [10], with permission).

In their recent study, cytochrome c could be loaded into large pore mesoporous silica nanoparticles at levels up to 415 mg/g particles and be released in a controlled manner, with full retention of activity, by varying the surrounding pH. The loaded particles were taken up by HeLa cancer cells, apparently by an endocytotic mechanism, escaped endosomal entrapment and released protein into the cytoplasm. The attractiveness of this approach for oral protein delivery resides in the high loading capacity, the protection offered against degradation in the gut, the ability to tune the particle characteristics for controlled release and their ability to cross the cell membrane and escape the endosomal compartment.

The safety of all non-biological systems for diagnostics and therapeutics is the focus of intense investigation especially when theses materials are fashioned in the nanoparticle form. Silica is no exception, even though amorphous silica is generally regarded as safe and is approved for use in food. Recently the cytotoxicity of silica nanoparticles has been examined in cell culture for a variety of human cell lines: normal dermal and pulmonary fibroblasts, as well as tumour cells of the colon, gastric system and lung [11]. Silica and silica-chitosan composite nanoparticles of 150-300 nM hydrodynamic diameter were incubated with cultured cells for 48 hours and their effect on cell viability/proliferation and membrane damage examined. The cytotoxicity of silica nanoparticles increased with increasing particle mass but became significant only at the high concentration of 667 μg/ml where viability was reduced to 60-80% of the control. Silica-chitosan composite nanoparticles were not cytotoxic even at these high concentrations. Depending, therefore, on the concentration and compositions used, mesoporous nanoparticles may prove to be effective devices in improving oral protein delivery.

Nanoparticle Formulations

There is substantial evidence both from pharmacological and from toxicological research that nanoparticles of specific composition, surface properties and size can efficiently pass the cell membrane. Polymeric carriers with formulations of varying compositions including hydrophilic/hydrophobic block co-polymers, poly anhydrides, poly (methyl methacrylate), poly alkyl cyanoacrylate, PLGA, PLG, alginate and dextran are being tested. One example of a successful nanoparticle formulation for oral delivery of peptides involves a totally new player, hydrophobin.

At Biomade Technology Foundation we have focussed on the use of this protein, hydrophobin-SC3, from the edible fungus *Schizopyllum commune* [12,13] for the formulation of nanoparticles of hydrophobic drugs. This small amphipathic protein of approximately 100 amino acids assembles at hydrophilic/hydrophobic interfaces such as oil/water or air/water interfaces, undergoes an irreversible conformational change leading to strongly associated molecules which appear as films on surfaces or as foams or emulsions. Depending on the processing conditions stable vesicles approximately 300 nM in size can be formed entrapping hydrophobic drugs. We have chosen cyclosporin, a hydrophobic cyclic peptide of very poor oral bioavailability to test hydrophobin vesicles as a potential delivery system. Formulations were prepared in ethanol/water mixtures and tested for uptake after oral administration in rats. Neoral, the commercial preparation of cyclosporin intended for oral administration, was taken along as the benchmark. The results (Figure 4) show essentially the same level of uptake with the hydrophobin nanocapsules as with Neoral; approximately 20% bioavailability in both cases. The hydrophobin formulation showed a prolonged uptake and a lower peak concentration resulting in a more constant, longer lasting drug level in the body [14]. Given the early stage of development further improvements in the hydrophobin-coupled uptake of cyclosporine are expected. Applications for the oral delivery of other hydrophobic peptides are also planned. The mechanism of this enhanced uptake is still under investigation; however, it should be different from Neoral since the Neoral formulation includes permeation enhancers while the hydrophobin formulation does not.

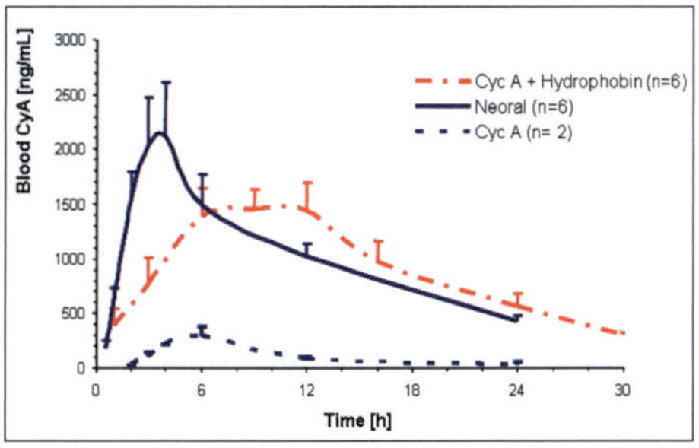

Figure 4. Cyclosporin A in the blood after oral administration. Concentrations determined by Elisa.

Receptor-Mediated Endocytosis

In receptor-mediated endocytosis, the recognition event is followed by formation of membrane structures that allow the uptake of individual molecules but also of larger structures such as macromolecular complexes but also nanoparticles and even microparticles. These larger structures function to protect proteins from degradation in the gut but also to enable the delivery of a therapeutically effective dose. Four examples of recent developments will be elaborated below.

Pseudomonas aeruginosa exotoxin A System

Publication of patent applications in the past two year have brought attention to a receptor-mediated endocytotic system being developed by Mrsny and colleagues with potential for applications in oral protein delivery [15,16]. The system is based on *Pseudomonas aeruginosa* exotoxin A, a 67 kDa protein composed of three distinct globular domains that is excreted by the bacterium. The N-terminal domain binds to the α2 macroglobulin receptor. The second domain is responsible for excretion of the protein and the third domain, for its cytotoxic activity. Deletion of the C-terminal portion of this domain detoxifies the protein. For the purpose of achieving oral protein delivery, genetic constructs have been made in which the receptor binding domain and the second domain, which is believed to be responsible for transcytosis through the cell following binding to the receptor, are retained but the cytotoxic domain is replaced by the therapeutic protein of interest. Furthermore, a sequence has been inserted into the linker between the first and second domains encoding a proteolytic cleavage site. The choice for a specific proteolytic cleavage was dictated by the goal of achieving cleavage as close as possible to the basolateral side of polarized epithelial cells, for a more efficient release of the cargo protein following transcytosis. Supporting data provided in the patent applications indicates that the system work efficiently in the uptake and transport of growth hormone through monolayer cultures of polarized human colon cancer Caco-2 cells as well as primary cultures of human tracheal epithelia and that there is selective cleavage and release of growth hormone from the basolateral side of the cells. Delivery of growth hormone with these constructs by oral gavage in mice was reported to result in high systemic serum levels as assessed by Elisa. That the orally delivered growth hormone in the serum samples possessed biological activity was verified *in vitro* by growth hormone-dependent cell proliferation assays. The bioavailability, when compared with subcutaneous injection was calculated to be surprisingly high, greater than 50%.

It should be noted that, in contrast to the nanoparticle-B_{12} system described in the next paragraphs, this system is designed to take therapeutic proteins across the cell one at a time so that achieving therapeutically effective doses may pose to be a challenge in some cases. Whether it can be used for uptake of nanoparticle has not been demonstrated although claims have been made that it is able to transport insulin aggregates. None of the data on this delivery system has yet appeared outside of patent applications making further confirmation necessary before focussing a lot of hope on it. Nevertheless, it does offer intriguing perspectives.

Vitamin B_{12} Uptake System

Vitamin B_{12}, when released from dietary components is bound first to haptocorrin, that is present in saliva and in the stomach. In the small intestine, pancreatic proteases digest the binding protein, releasing vitamin B_{12} which then becomes bound to Intrinsic Factor, a glycoprotein of approximately 45 kDa with a 10^{10} affinity constant for B_{12}. Receptors for Intrinsic Factor on the ileal mucosa bind the complex and undergo endocytosis, allowing vitamin B_{12} to be absorbed into portal blood. Since the mid-90s this system has been the object of several attempts to piggy-back peptide cargos into circulation, the concept being that the binding proteins in the stomach and Intrinsic Factor itself would offer some protection to the cargo from proteolytic degradation in the gut. A major limitation with the early approach of direct attachment of B_{12} to the cargo protein was the low uptake capacity of the system which prevented achieving therapeutically effective levels. More recently, the concept has been expanded to nanoparticles which serve the dual role of protecting the proteins incorporated in the particle matrix from degradation and of delivering a therapeutically relevant amount by transporting many molecules per endocytotic event [17,18]. The matrices have been constructed from a variety of polymers, decorated with B_{12} linked through spacers and loaded with protein. An example of a cross-linked dextran nanoparticle system developed and tested for oral insulin delivery is shown in Figure 5. Particles, 150-250 nM in diameter, were produced containing 4% (w/w) B_{12} and 2-4% (w/w) insulin with a loading efficiency of 50-70%. Upon oral administration in animals a rapid, extended hypoglycaemic effect was found with a maximum pharmacological availability of 26% [17]. Furthermore, particles not carrying the B_{12} ligand gave a statistically poorer result supporting the concept that uptake occurred via the B_{12}-IF receptor system as designed.

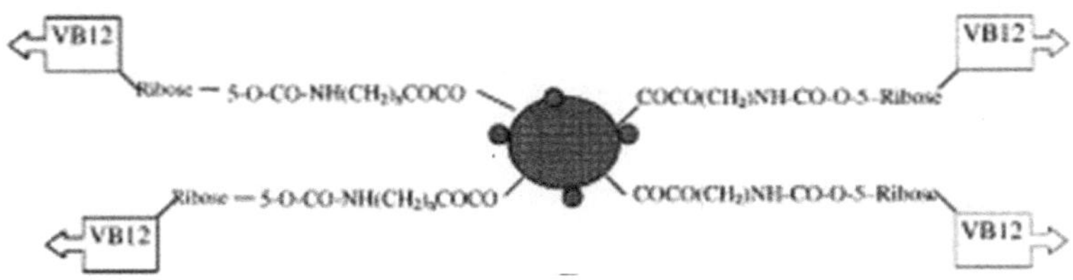

Figure 5. Reproduced from reference [17], with permission.

Whether or not the nanoparticle system is essential for achieving therapeutic efficacy is challenged by recent results of Petrus *et al* [19] who linked B_{12} through its ribose moiety directly to a single lysine residue on the insulin B strand and were able to achieve a sustained 5 hour hypoglycaemic response in rats in response to oral administration. A 10,000-fold excess of free B_{12} blocked the effect indicating that the insulin construct was gaining entry via the B_{12} uptake pathway.

Transferrin Receptor-Mediated Uptake

Transferrin is a very high affinity iron-chelating glycoprotein of about 80 kDa which transports iron into the blood though the transferrin receptor-mediated endocytotic pathway. High levels of the receptor are present on brain and GI tract endothelial cell layers. The stability of transferrin against degradation by proteolytic enzymes in the gut and the high level of receptors have made the system a favourite candidate for enhancing the uptake of therapeutic agents including proteins [20-22]. An apparent drawback of the system is that the majority of transferrin receptors are located on the basolateral side of polarized cells and have only a short residence time on the apical side, thus limiting their availability for binding proteins targeted from the gut. Nevertheless Lim & Shen [23] recently showed that aggregated transferrin, cross-linked to an average of 3.5 molecules per complex, was more efficiently transferred across the epitheium than monomeric transferrin presumably due to alterations in the transcytosis pathway used by the two forms. It was also more stable to proteolytic degradation by gut proteases than the monomeric protein and had a longer residence time in the blood. All three effects carried through to the insulin-coupled forms and resulted in a higher efficacy of the aggregated transferrin-insulin complex at reducing blood sugar levels compared to the monomeric construct or free insulin in experimental animals.

Thus the nano-particulate nature of the aggregated transferrin-insulin complex has a positive effect on both the proteolytic stability of the therapeutic protein in the gut and the efficiency of transfer from the gut to the blood and a longer residence time in the blood.

Polymeric Immunoglobulin Receptor System

The challenge of achieving a therapeutically effective dose of a peptide or protein might be met even more easily using the Polymeric Immunoglobulin Receptor system (pIgR) because of its ability to transcytose not just immunoglobulins but whole bacteria much larger than nanoparticles. This receptor, which plays a key role in mucosal immunity, is an integral membrane protein responsible for the transport of IgA into the mucosal layer [24-26]. IgA, produced in the lamina propria underlying the epithelial cell layer, binds in the dimeric form to the extracellular domain of the receptor and is transported through the epithelial cell to the apical layer. Proteolytic cleavage of the extracellular domain from the transmembrane domain occurs, releasing IgA and the extracellular domain together into the mucosal layer as secretory immunoglobulin, S-IgA.

The receptor is known to also cycle to the apical surface without carrying IgA so that its extracellular domain remains attached. Studies to unravel the process of entry of pathogens through the nasopharygeal epithelium have shown that they make use of these intact receptors, binding to them and following the reverse route through the epithelial cell back to the basolateral side [26]. This was demonstrated with monolayers of fully polarized hpIgR-transfected MDCK cells grown on Transwell membranes. The membranes took up 15x more bacteria than those first exposed to antiserum against the extracellular domain, indicating that a significant quantity of functional pIgR is present at the apical side. Furthermore transmigration from the apical to the basolateral side was shown to occur with the Detroit 562 human cell line without harming the integrity of the monolayer. Transmigration was indicated by the number of bacteria which ended up in the lower chamber which was nine times higher than the controls.

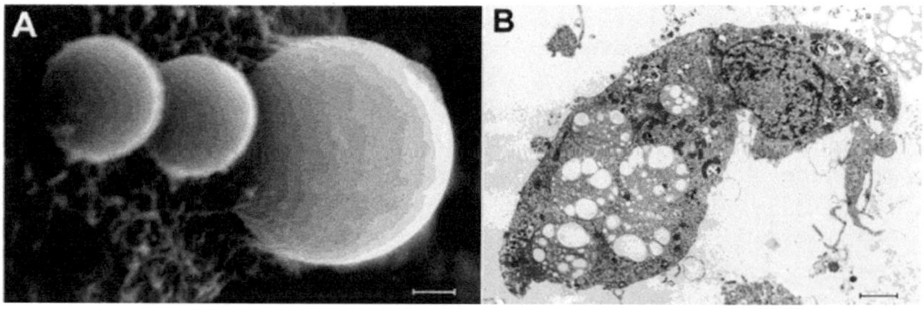

Figure 6. Uptake of CbpA1-coated beads by Detroit-cells: A) Coated beads. Bar is 0.8 μm. B) Detroit-cells with internalized CbpA1-coated beads, TEM of ultra thin section. Bar is 1.6 μm (reproduced from reference [26], with permission).

Pathogens attach to the receptor extracellular domain through proteins on their surface. The interaction of one such protein, Choline binding protein A (CbpA) of *Streptococcus pneumonia,* has been characterized in detail. Antibodies against the isolated extracellular domain prevent its binding to purified CbpA. Mutants lacking CbpA can also not bind to the receptor. An experimental result which offers intriguing perspective for the use of the pIgR system for oral protein delivery is shown in Figure 6. Here micron-size latex beads coated with CbpA1, a truncated portion of CbpA, containing just the two choline binding repeat domains were internalized by Detroit 562 human cell line through the mediation of pIgR. The size of these spheres, and the observation that bacteria can transmigrate across a monolayer of cells in culture offers promise for being able to deliver packages containing high doses of therapeutic proteins or peptides via oral administration.

Conclusion

Oral delivery of proteins is a complex challenge. Some of the issues were summarized at the beginning of this paper. There are realistic contributions which nanotechnology can provide, especially in the area of protection against degradation from the harsh environment in the gut and achieving a therapeutically relevant dose. It is unrealistic to expect that the bioavailability via oral administration will be able to match that achieved by direct injection but the additional cost might be justified in the light of better patience compliance and quality of life which oral administration can offer.

References

[1] http://www.uspto.gov/patft/.

[2] Nanosciences and Nanotechnologies, opportunities and uncertainties: ISBN 0 85403 604 0 (2004). The Science Policy Section, The Royal Society, London SW1Y 5AG, UK. (http://www.nanotec.org.uk/report/Nano%20report%202004%20fin.pdf).

[3] M.C. Roco. National Nanotechnology Initiative - Past, Present, Future. In: W.A. Goddard III, D.W. Brenner, S.E. Lyshevski (Eds): *Handbook of Nanoscience, Engineering, and Technology* (2nd Ed). Taylor and Francis (2007) 3.1-3.26.

[4] A. Ahmed, C. Bonner, T.A. Desai. Bioadhesive microdevices with multiple reservoirs: a new platform for oral delivery. *J. Control. Release* **81** (2002) 291-306.

[5] S.L. Tao and T.A. Desai. Microfabricated drug delivery systems: from particles to pores. *Adv. Drug Deliv. Rev.* **55** (2003) 315-328.

[6] S.L. Tao and T.A. Desai. Micromachined devices: the impact of controlled geometry from cell-targeting to bioavailability. *J. Control. Release* **109** (2005) 127-138.

[7] J.S. Beck, J.C. Vartuli, W.J. Roth *et al.* A new family of mesoporous molecular sieves prepared with liquid crystal templates. *J. Am. Chem. Soc.* **114** (1992) 10834-10843.

[8] C.T. Kresge, M.E. Leonowicz, W.J. Roth *et al.* Ordered mesoporous molecular sieves synthesized by a liquid-crystal template mechanism. *Nature* **359** (1992) 710-712.

[9] B.G. Trewyn, C.M. Whitman, V.S.-Y. Lin. Morphological control of room-temperature ionic liquid templated mesoporous silica nanoparticles for controlled release of antibacterial agents. *Nano Letters* **4** (2004) 2139-2144.

[10] I.I. Slowing, B.G. Trewyn, V.S.-Y. Lin. Mesoporous silica nanoparticles for intracellular delivery of membrane-impermeable proteins. *J. Am. Chem. Soc.* **129** (2007) 8845-8849.

[11] J.S. Chang, K.L. Chang, D.F. Hwang *et al.* In vitro cytotoxicity of silica nanoparticles at high concentrations strongly depends on the metabolic activity type of the cell line. *Environ. Sci. Technol.* **41** (2007) 2064-2068.

[12] K. Scholtmeijer, J.G. Wessels, H.A. Wösten. Fungal hydrophobins in medical and technical applications. *Appl. Microbiol. Biotechnol.* **56** (2001) 1-8.

[13] M.L. de Vocht, I. Reviakine, W.P. Ulrich *et al.* Self-assembly of the hydrophobin SC3 proceeds via two structural intermediates. *Protein Sci.* **11** (2002) 1199-1205.

[14] A. van den Berg, M. Haas, R. Rink *et al.* Peptide drug formulations with hydrophobins, and uses thereof. Patent application (2008).

[15] R.J. Mrsny. Methods and compositions for needleless delivery of binding partners. US patent application; US 2007/0148131 A1 (2006).

[16] R.J. Mrsny. Methods and compositions for needleless delivery of macromolecules. US patent application; US 2006/0153798 A1 (2005).

[17] K.B. Chalasani, G.J. Russell-Jones, S.K. Yandrapu *et al.* A novel vitamin B_{12}-nanosphere conjugate carrier system for peroral delivery of insulin. *J. Control. Release* **117** (2007) 421-429.

[18] K.B. Chalasani, P.V. Diwan, K.V. Rughavan *et al.* Vitamin B_{12} biodegradable microparticulate conjugate carrier systems for peroral delivery of drugs, therapeutic peptides/proteins and vaccines. U.S. patent 6,482,413 (2002).

[19] A.K. Petrus, A.R. Vortherms, T.J. Fairchild *et al.* Vitamin B_{12} as a carrier for the oral delivery of insulin. *ChemMedChem.* **2** (2007) 1717-1721.

[20] U. Bickel, T. Yoshikawa, W.M. Pardridge. Delivery of peptides and proteins through the blood-brain barrier. *Adv. Drug. Deliv. Rev.* **46** (2001) 247-279.

[21] Z.M. Qian, H. Li, H. Sun *et al.* Targeted drug delivery via the transferrin receptor-mediated endocytosis pathway. *Pharmacol. Rev.* **54** (2002) 561-587.

[22] H. Li and Z.M. Qian. Transferrin/transferrin-receptor-mediated drug delivery. *Med. Res. Rev.* **22** (2002) 225-250.

[23] C.J. Lim and W.C. Shen. Comparison of monomeric and oligomeric transferrin as potential carrier in oral delivery of protein drugs. *J. Control. Release* **106** (2005) 273-286.

[24] K.E. Mostov and G. Blobel. A transmembrane precursor of secretory component: the receptor for transcellular transport of polymeric immunoglobulins. *J. Biol. Chem.* **257** (1982) 11816-11821.

[25] K.E. Mostov and C.S. Kaetzel. Immunoglobulin transport and the polymeric immunoglobulin receptor. In: P.L. Ogra, J. Mestecky, M.E. Lamm *et al.* (Eds): *Mucosal Immunology* (2nd Ed). San Diego, Academic Press (1999) 181-211.

[26] J.R. Zhang, K.E. Mostov, M.E. Lamm *et al.* The polymeric immunoglobulin receptor translocates pneumococci across human nasopharyngeal epithelial cells. *Cell* **102** (2000) 827-837.

Towards Drugs of the Future
C.G. Kruse and H. Timmerman (Eds.)
IOS Press, 2008

doi:10.3233/978-1-58603-949-3-101

The Omeprazole Story: How Endurance in Research May Pay Off

Enar Carlsson
Nordreviksvägen 23A, 42737 Billdal, Sweden

Abstract. Based on the dictum: No acid - no ulcer a number of pharmaceutical companies started up research aiming at a drug, inhibiting gastric acid secretion during 1960s. The focus of that research was gastrin, the gastric acid stimulating hormone released from G-cells in the antral part of the stomach. The Searle company in the US tried to find a small molecule inhibiting the gastrin receptor, while AstraHässle in Sweden tried to find something that decreased the release of gastrin from the G-cells. SK&F focused on the H_2-receptor also involved in the regulation of gastric acid secretion. The SK&F approach lead to the first clinically useful acid inhibitor, cimetidine. The AstraHässle approach was a dead end, as was the Searle approach. However, the Searle idea led to the discovery of CMN 131, a compound which had some acid inhibitory effect, but was also toxic. When AstraHässle failed with their first attempt, they picked up on CMN 131 as a new lead around 1972. This approach resulted in timoprazole (1974), picoprazole (1977), and eventually omeprazole (1979). Omeprazole was first launched in Sweden 1988, under the trade name Losec. In 1997 Losec reached a yearly sale of more that 6 billion USD and became the world's biggest drug.

The AstraHässle project, which ended up with the approval of omeprazole 1988, started already 1967. Thus it took more than 20 years from the idea to market. Would that have been possible today with the present time-press and recourse restrictions within drug industry? It's doubtful. The struggle and issues to be solved during the journey are described.

Keywords. Omeprazole, Losec, CMN 131, development issues

Almost 100 years ago the Croatian physiologist Karl Schwarz formulated the dictum: No acid - no ulcer. Thus, it has been known for a long time that gastric acid plays an important role for peptic ulcer disease. However, it took until the 1960s before pharmaceutical companies made severe attempts to develop effective inhibitors of gastric acid secretion. Until then, milk regiments, diet restrictions and antacids in addition to different surgical procedures were used to treat peptic ulcer disease.

The discovery by Black and co-workers of the H_2-receptors and their role in the regulation of gastric acid secretion led to the development of selective H_2-receptor antagonists like cimetedine and ranitidine. These drugs made the difference and meant a real breakthrough in the treatment of acid related diseases [1].

In parallel, other companies started up research aiming at an inhibitor of gastric acid secretion based on the recently gained knowledge of the origin, role and structure of gastrin, the gastric acid secretion stimulating hormone. The Searle & Co company in the US tried to find a small molecule with affinity and antagonistic properties on the gastrin receptor. Searle utilised the C-terminal of gastrin as the lead structure (Figure 1). A molecule SC-15396 "antigastrin" was identified with some, but too weak, antisecretory properties [2]. Later on, the Servier company in France picked up on the idea from Searle and tried to modify the "antigastrin" molecule. Servier's attempt resulted in CMN 131, which seemed to be slightly more potent than "antigastrin" but showed acute toxicity hindering further development [3]. An interesting serendipity observation was made in the Searle laboratories, which lead to the development of the sweetener Aspartame. The relation of the chemical structure of Aspartame and the C-terminal of gastrin is obvious (Figure 1).

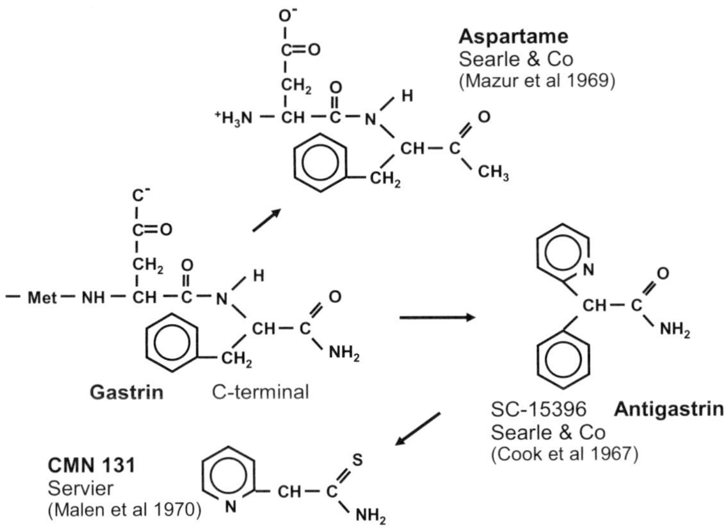

Figure 1. From the C-terminal of gastrin to antigastrin, CMN 131 and Aspartame.

The Gastrin Project 1967

At Hässle, the research company outside Gothenburg belonging to the Swedish drug company AB ASTRA, the idea came up that acid secretion could be decreased by blocking the release of gastrin from the G-cells in the antral part of the stomach. Some data originating from the Karolinska Institute indicated that such an effect could be achieved by oral administration of a local anaesthetic like xylocaine. The aim was therefore to develop a non-absorbable, locally acting anaesthetic. The first candidate drug (CD) H 81/75 selected and tested in man in 1970 was without any antisecretory effect in humans, in spite of the fact that it was quite effective *in vivo* in rats [4]. A new idea and a new lead were needed.

From CMN 131 to Omeprazole

On January 4th 1979 the Hässle chemist Ylva Örtengren described on page 68 in her laboratory notebook nr H 168 the synthesis of a compound receiving the code number H 168/68 (Figure 2). This compound later on got the generic name omeprazole and the tradename Losec® (in most countries). In 1997 Losec® became the best selling pharmaceutical in the world reaching a yearly sale of more than 6 billion USD at peak.

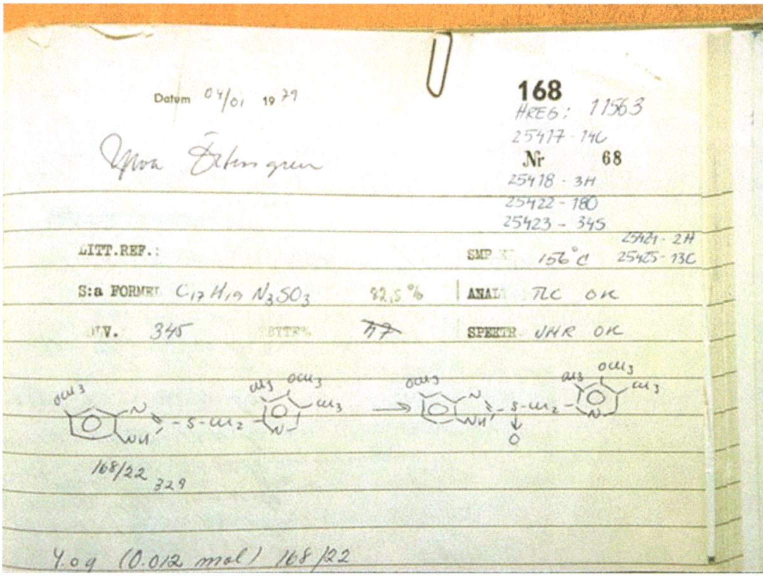

Figure 2. H 168/68 was synthesized on the 4th of January 1979. From Ylva Örtengren's notebook.

Milestones

• Start of project:	1967	
		~12 years
• Synthesis of H 168/68:	1979	
		~ 9 years
• Approval and launch in Sweden:	1988	
		~ 8 years
• World leader:	1996	

Use and sales

- 1 billion treatment courses
- Peak yearly sales >6 billion USD
- Added sales >50 billion USD

Figure 3. The Omeprazole story in short.

Figure 3 gives the Losec story in a nut shell. The molecule omeprazole was synthesized in 1979 i.e. 12 years after start of the original gastrin project and was approved in 1988 after more than 20 years of research and development. Today more than 1 billion treatment courses (which can be compared to the whole population in China) have been produced and distributed. The added sales amount to above 50 billion USD. The discovery and development of omeprazole is an almost unprecedented success story within drug industry [4-7].

However, the long research and development time as such indicates that the project did not follow a straight, paved road without hurdles.

Figure 4 gives project milestones from the start 1967 until the approval and launch of the follow-up compound esomeprazole. As already mentioned above, the first CD

H 81/75 belonging to the initial compounds with a local anaesthetic mechanism of action, failed to give any effect on acid secretion when tested in man.

1967	Start of the "Gastrin project"
1970	No effect of H 81/75 in man
1971	Presentation of CMN 131
1974	H 83/69 Timoprazole (**thyrotoxicity**)
1977	H 149/94 Picoprazole (**vasculitis**)
1979	Picoprazole given to man
1979	***Omeprazole synthetized***
1982	Symposium OMGE, Stockholm
1984	Clinical trials suspended (**carcinoids in rats**)
1988	Approval in Sweden
1990	1 million treatments
1990	**The Genotox story**
1993	50 million treatments
1994	100 million treatments
1995	**The BGA story**
1997	Worldleader
2000	***Approval of Esomeprazole***

Figure 4. Milestones 1967 - 2000.

New ideas had to be sought and came when CMN 131 was presented at a pharmacological congress in Hungary in 1971 [3]. CMN 131 was a potent inhibitor of gastric acid secretion in an animal model but its acute toxic effect hindered its further development. The Hässle chemists, however, had ideas on how to get rid of the toxicological problems.

The first CD based on the new lead compound was H 124/26 or rather its active metabolite H 83/69, timoprazole, was tested in toxicological studies. The pathology examination revealed, however, enlargement of the thyroid gland of treated rats. Mechanistic studies showed that this thyrotoxic effect was related to inhibition of iodine uptake in the gland. H 83/69 could not be tested in man. There were reasons, though, to believe that this effect on iodine uptake was not related to the primary pharmacological effect on acid secretion [4]

Subsequent compounds were therefore tested both for inhibition of iodine uptake in the thyroid and inhibition of gastric acid secretion. The two pharmacological effects could be separated. At a specific range of lipophilicity of the compounds the inhibitory effect on acid secretion was maximized while within this window of lipophilicity the iodine uptake was minimized or absent [4].

A couple of new compounds, among them H 149/94, were selected as CDs 1977. Both of them were lacking interference with iodine uptake in the thyroid but with potent acid antisecretory properties in the stomach. Toxicological studies were performed in rats and dogs.

This time, none of the compounds showed thyrotoxicity. But another serious toxicological effect was revealed. Some treated dogs were found to have necrotizing vasculitis in the small intestine, again stopping tests in man. Astra management, as had happened before, required now termination of all further work in the project. Ten years without a CD in development was enough [4].

But again an explanation for the toxicological problem was found. A detailed analysis of the dogs included in the studies revealed that one dog in the non-treated control group also had developed necrotizing vasculitis. Furthermore this control dog had the same father, Fabian, as all dogs with vasculitis among the treated dogs. At the time it was believed that the vasculitis was a hypersensitivity reaction to fragments of intestinal worms as a result of the antiparasitic treatment in genetically predisposed dogs.

Today we know that Fabian and his offspring carried hereditariness for the Beagle Pain Syndrome, which is associated with necrotizing vasculitis in various vascular beds. This hidden disease in beagle dogs can be activated in various ways including stress and chemicals.

After repeating the toxicity study in dogs free from this hereditary problem, H 149/94 could be cleared and given to man.

Figure 5 shows the effect of an intraduodenally given dose of H 149/94 to a Zollinger.Ellison (ZE) patient with massive hypersecretion of gastric acid. Impressive long acting, antisecretory effect of a single dose of H 149/94 was demonstrated. This happened in 1979 at the same time as the compound H 168/68 had been synthesized and found to be an even more potent acid inhibitor than H 149/94.

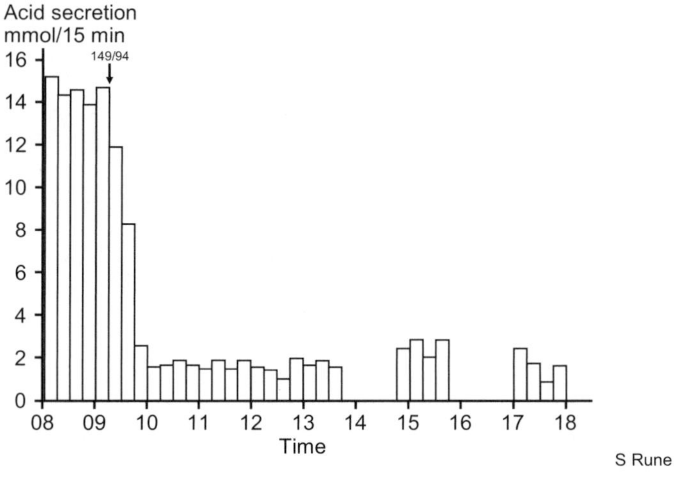

Figure 5. Effect of H 149/94, picoprazole on gastric acid secretion in a Zollinger.Ellison patient treated by Professor Simon Rune, Copenhagen.

Which compound should be chosen for further development? H 149/94 had been found to be active in man, but the vasculitis in the first dog toxicity study would always have to be explained. H 168/68 was more potent and so far without any shadows in form of toxicity. Indeed a delicate decision problem.

H 168/68, omeprazole was chosen and a crash development program was designed.

Omeprazole was found to be the most potent inhibitor of gastric acid secretion in rats and dogs *in vivo* [8]. There were no thyrotoxicity in the rat, no necrotizing vasculitis in the dog and no other signs of toxicity at reasonable doses were found in the initial safety studies. An Investigational New Drug (IND) application was filed in 1980.

The first and highly encouraging, clinical efficacy data was reported at the OMGE congress in Stockholm in 1982. Out of 26 duodenal ulcer patients 25 were healed after 4 weeks treatment with 40 mg orally. Over the next couple of years there was full speed ahead with the developmental program, aiming at submission of the New Drug Application (NDA) in 1985.

Further Toxicological Problems

In May 1984 only one safety study remained to be reported for the NDA, the lifelong carcinogenicity study in rats. Clinical trials were ongoing worldwide when reports arrived from the toxicological contract laboratory in the USA, performing the study. Gastric endocrine tumours (carcinoids) had been found in a significant number of omeprazole treated rats (Figure 6). The frequency of carcinoids was clearly both treatment- and dose-dependent and there was no free dose group [9]. Within one week, all ongoing clinical studies had been stopped, except for a few in which patients with life threatening disease like the ZE syndrome and very severe reflux eosophagitis were treated.

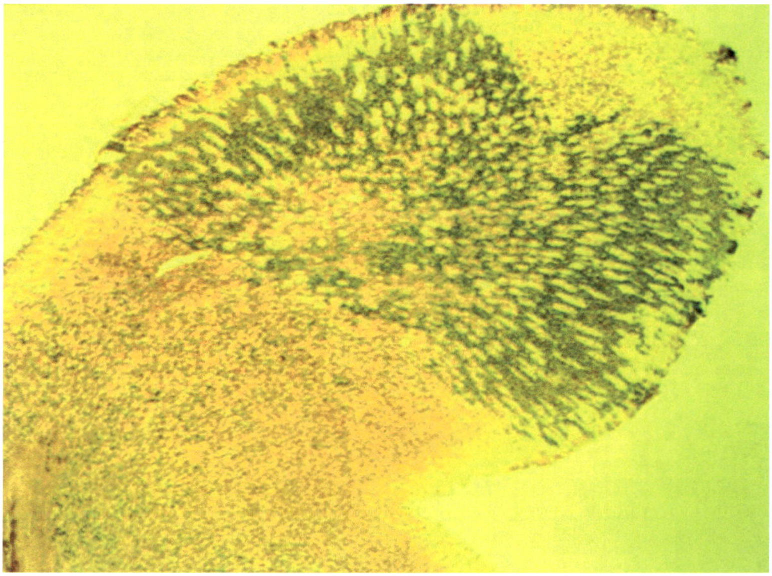

Figure 6. Focal ECL-cell carcinoid in the gastric mucosa of a rat treated with a high oral dose of omeprazole for 2 years. Sevier-Munger silver-staining.

The carcinoides originated from enterochromaffine-like (ECL) cells, a type of endocrine cells only located in the gastric mucosa. The ECL-cells synthesize and secrete histamine in response to gastrin. In addition to this effect, gastrin has also a potent trophic action on the ECL-cells. Combining this with the fact that almost complete inhibition of acid secretion with omeprazole leads to massive release of gastrin and hypergastrinaemia in treated rats, suggested a possible explanation for the observed gastric endocrine tumours in the lifelong rat study.

This mechanism was demonstrated to be the cause of the ECL-cell hyperplasia in omeprazole-treated rats, as the hyperplasia did not occur in treated rats subjected to resection of the gastric antrum [10]. Later studies could also demonstrate that ECL-cell carcinoids developed in rats treated lifelong with the H_2-receptor antagonist, ranitidine [11], as well as by partial gastric corpectomy also resulting in massive hypergastrinaemia [12].

After about one year, in the beginning of 1985 sufficient documentation supporting the "gastrin hypothesis" (Figure 7) had been gathered and the regulatory authorities allowed restart of the clinical studies.

- Rats display high ECL-cell density and spontaneous ECL cell hyperplasia in the oxyntic mucosa
- Elevated, sustained gastrin levels in rats result in augmented ECL-cell hyperplasia and, at the end of life, gastric carcinoid formation

Short-term studies	**Life-long studies**
ECL-cell hyperplasia carcinoids	**ECL-cell hyperplasia and**
– acid inhibition	– acid inhibition
– proton pump blockade	– proton pump blockade
– H_2-blockade	– H_2-blockade
– gastrin administration	
– partial fundectomy	– partial fundectomy

- ECL-cell hyperplasia is fully reversible following normalization of the gastrin levels

Figure 7. The gastrin mechanism for ECL-cell hyperplasia and carcinoid development. List of supporting studies.

Efficacy in Clinical Studies

Omeprazole was found to be significantly superior to previous treatment with H_2-receptors antagonists in patients with duodenal and gastric ulcers. A particular notable superiority of omeprazole compared with H_2-receptor antagonist was found in GERD patients, in which healing rates were about twice as high with omeprazole [13].

Omeprazole was registered and launched in Sweden in 1988, about 20 years after start of the project in 1967.

Mechanism of Action

The mechanism of action is summarized in Figure 8. The molecule omeprazole has several characteristics that are important for its mechanism of action. First, omeprazole is highly lipophilic, which means that it easily penetrates cell membranes. Second, it is a weak base, which means that it concentrates in acid compartments. Third, it is very unstable in an acidic milieu. The half-life is about 2 minutes at pH 2, whereas at pH 7.4 it is about 20 hours. Omeprazole is a prodrug that accumulates within the acid secretory cannaliculi of the target parietal cell where it is transformed to the active inhibitor, the sulphenamide. The sulphenamide binds to and blocks the activity of the proton-transporting H+K+ATPase in the secretory membrane of the parietal cell.

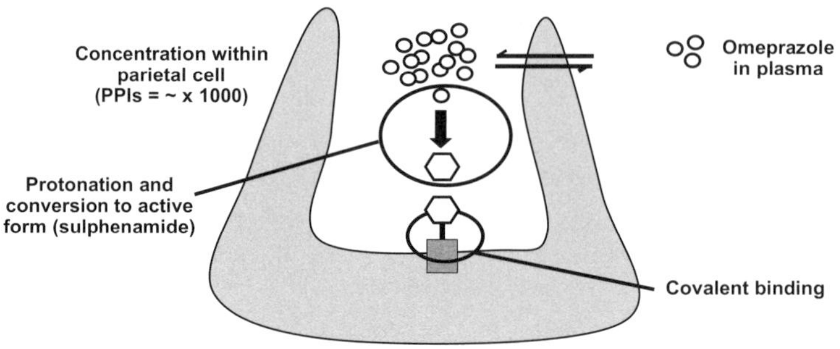

Figure 8. Mechanism of action of omeprazole.

Reactivation of the acid producing capacity of the parietal cell is partly due to the effect of endogenous glutathione breaking the bond between the sulphenamide and the enzyme, partly to synthesis of new enzyme molecules. This results in a long duration of effect with an apparent half-life of 24 hours of the antagonist/enzyme complex in contrast to the short plasma half-life, 1-2 hours of omeprazole [5-7,14].

Rapid Growth in Spite of Serious Attacks

The most important milestones of the Losec® project are shown in Figure 4. Already a couple of years after the first launch in 1988, 1 million treatment courses had been given. In 1994 the corresponding figure was 100 million. Today more than 1 billion treatment courses have been given, which can be compared with the whole population in China. In 1997 omeprazole became the best selling drug in the world reaching about 6 billion USD in yearly sales. The added sales amount to more than 50 billion USD. In year 2000 the follow-up compound esomeprazole was launched.

This development after launch has not happened without attacks needing mobilization of large parts of the company including preclinical and clinical research, regulatory and safety. In 1990 a competitor company published data in the Lancet interpreted as a genotoxic effect of omeprazole. Internal as well as external studies could demonstrate that the method used in the competitor study was flawed and that omeprazole had no genotoxic properties. However, as a rule one such negative study needs ten studies to take away the suspicion. Regarding the carcinoids it took ten years and many, many studies to convince doctors that there was no risk to use omeprazole even long-term.

In 1994 a consumer magazine in Germany published single case reports on eye damage and even blindness in patients treated with i.v. omeprazole. The patients were seriously burned patients that had been kept anaesthetized for about a month and during this time had been treated with up to 30 different medications, among them i.v. omeprazole This was picked up by BGA finally requiring withdrawal of the i.v. formulation and including a warning text in the labelling. After thorough analysis of all safety documentation from clinical studies as well as data from post-launch clinical use, new studies and detailed analysis of all individual cases from Germany by world expertise it was concluded that there were no indications of a toxic effect on the eye. The european CPMP at a hearing in Brussels, also concluded that there was no reason to believe that omeprazole had anything to do with this eye problem in the seriously ill patients in Germany. All cases had been through a hypotensive crisis which is known to cause anoxic damage of eye nerves.

Extensive work, complete transparency and continuous contacts with regulatory authorities worldwide were all important components for the solution of this serious issue.

Things to Consider, Lessons to Learn?

The preclinical research project resulting in the synthesis of omeprazole took place at Hässle, a relatively small and autonomic research unit within AB ASTRA during the 1970s. Hässle was also responsible for the clinical development during the 1980s. This does not mean that Astra was without control, the overall research budget was decided by Astra. Within the budget frames, however Hässle could prioritize between the different cardiovascular and gastrointestinal projects and take decisions in research matters. There was a minimum of bureaucracy and a short distance between management and people in the research projects. Project leaders had also a high degree of empowerment. Research goals were based on clear medical needs and were therefore experienced as stimulating and lead to engaged scientists with unique performance.

An important success factor was the culture of openness and trust within the Hässle organization, which meant that specialists could be challenged. An issue was considered a common problem owned by everybody and should therefore be solved in collaboration. Furthermore, a problem, most often of toxicological character, should be understood before a go/no go decision was taken.

A consequence of this was also that it was easy to "mobilize" the company in crisis situations. The Genotox and BGA issues are good examples of that.

Would it be possible to develop omeprazole in today's pharmaceutical research environment? The methodology today with robotisized *in vitro* HTS is not suited to identify a prodrug like omeprazole. For that, close to *in vivo* methodology is required.

The time pressure and requirement to deliver within defined periods of time and the bureaucracy within Big Pharma are other hurdles which are higher today compared to the situation 20-30 years ago.

The question can be debated.

References

[1] C.R. Ganellin. Development of anti-ulcer H_2-receptor histamine antagonists. In: J. Fischer and C.R. Ganellin (Eds): *Analogue-based drug discovery*. Wiley-VCH Verlag Weinheim (2006) 71-80.

[2] D.L. Cook and R.G. Bianchi. SC-15396: a new antiulcer compound possessing anti-gastrin activity. *Life Sci.* **6** (1967) 1381-1387.

[3] C.E. Malen and B.H. Danree. New thiocarboxamides derivatives with specific gastric antisecretory properties. *J. Med. Chem.* **14** (1971) 244-246.

[4] S.E. Sjöstrand, L. Olbe, E. Fellenius. The discovery and development of the proton pump inhibitor. In: L. Olbe (Eds): *Proton pump inhibitors*. Birkhäuser Verlag Basel-Boston-Berlin (1999) 3-20.

[5] E. Carlsson, P. Lindberg *et al.* "Two of a kind". *Chem. Brit.* **38** (2002) 42-45.

[6] L. Olbe, E. Carlsson, P. Lindberg. A proton-pump inhibitor expedition: the case histories of omeprazole and esomeprazole. *Nat. Rev. Drug Discov.* **2** (2003) 132-139.

[7] P. Lindberg and E. Carlsson. Esomeprazole in the framework of proton-pump inhibitor development. In: J. Fischer and C.R. Ganellin (Eds): *Analogue-based drug discovery*. Wiley-VCH Verlag Weinheim (2006) 81-113.

[8] H. Larsson, E. Carlsson, U. Junggren *et al.* Inhibition of gastric acid secretion by omeprazole in the dog and rat. *Gastroenterology* **85** (1983) 900-907.

[9] L. Ekman, E. Hansson, N. Havu *et al.* Toxicological studies on omeprazole. *Scand. J. Gastroenterol. Suppl.* **108** (1985) 53-69.
[10] H. Larsson, E. Carlsson, H. Mattsson *et al.* Plasma gastrin and gastric enterochromaffinlike cell activation and proliferation. Studies with omeprazole and ranitidine in intact and antrectomized rats. *Gastroenterology* **90** (1986) 391-399.
[11] N. Havu, H. Mattsson, L. Ekman *et al.* Enterochromaffin-like cell carcinoids in the rat gastric mucosa following long-term administration of ranitidine. *Digestion* **45** (1990) 189-195.
[12] H. Mattson, N. Havu, J. Bräutigam *et al.* Partial gastric corpectomy results in hypergastrinemia and development of gastric enterochromaffinlike-cell carcinoids in the rat. *Gastroenterology* **100** (1991) 311-319.
[13] L. Lundell and J. Dent. Gastro-oesophageal reflux disease. In: L. Olbe (Eds): *Proton Pump Inhibitors*. Birkhäuser Verlag Basel-Boston-Berlin (1999) 223-235.
[14] B. Wallmark and G. Sachs. Inhibition of the gastric proton pump. In: L. Olbe (Eds): *Proton Pump Inhibitors*. Birkhäuser Verlag Basel-Boston-Berlin (1999) 47-59.

Towards Drugs of the Future
C.G. Kruse and H. Timmerman (Eds.)
IOS Press, 2008

doi:10.3233/978-1-58603-949-3-111

Conceptual Design of Cannabinoid CB_1 Receptor Antagonists

J.H.M. Lange
Solvay Pharmaceuticals Research Laboratories, C.J. van Houtenlaan 36, 1381 CP Weesp, The Netherlands

Abstract. Cannabinoids constitute an area of intensive research, both in industry and in academic institutions. Approximately 15 years ago, tremendous progress has been made in the molecular characterization of endogenous cannabinoids and their receptors. Cannabinoid CB_1 receptor antagonists showed clinical efficacy in the treatment of obesity and improved cardiovascular and metabolic risk factors. They have good prospects in other therapeutic areas, including smoking and alcohol addiction. Solvay's research achievements in the fast-moving field of CB_1 receptor antagonists are highlighted in relation with the general state of the art. Solvay pursued several medicinal chemistry strategies. The application of the concept of conformational constraint led to the discovery of rigidified analogs of the prototypic CB_1 receptor antagonist rimonabant. Modifications at the 4-position of the pyrazole ring in rimonabant led to a novel compound with retained CB_1 receptor antagonistic potency but with a different predicted biodegradation profile. Bioisosteric replacement of the central heterocyclic pyrazole ring in rimonabant yielded imidazoles, triazoles and thiazoles as selective CB_1 receptor antagonists. Dedicated medium throughput screening efforts delivered one diarylpyrazoline hit. This initial hit showed poor pharmacokinetic properties but could be successfully optimized into the orally active and highly CB_1/CB_2 receptor selective drug candidate ibipinabant, which is the subject of a development agreement between Solvay and Bristol-Myers Squibb.

Keywords. Cannabinoid CB_1 receptor antagonist, inverse agonist, obesity, addiction, receptor selectivity, conformational constraints, bioisosterism, hydrogen-bond acceptor interaction, lipophilicity, oral bioavailability

Introduction

The cannabinoid research area is of interest for both the industry and academic institutions. Cannabinoids from the *Cannabis sativa L.* plant have been used for many centuries [1] as medicinal agents as well as for recreational purposes. In 1964, Mechoulam and his co-workers elucidated the chemical structure of the main psychoactive component [2] from hasjish, Δ^9-tetrahydrocannabinol (Δ^9-THC (**1**), Figure 1) [3]. Interestingly, Δ^9-THC (dronabinol) is nowadays being marketed by Solvay as Marinol® for the treatment of AIDS-related weight loss and to combat cachexia and emesis in cancer patients. The synthetic THC analogue nabilone (**2**) has also been approved in the USA whereas the standardized cannabis extract Sativex® (an oromucosal spray which contains Δ^9-THC and the non-psychotropic cannabidiol (**3**)) [4] has been approved in Canada for use in neuropathic pain in multiple sclerosis [5,6].

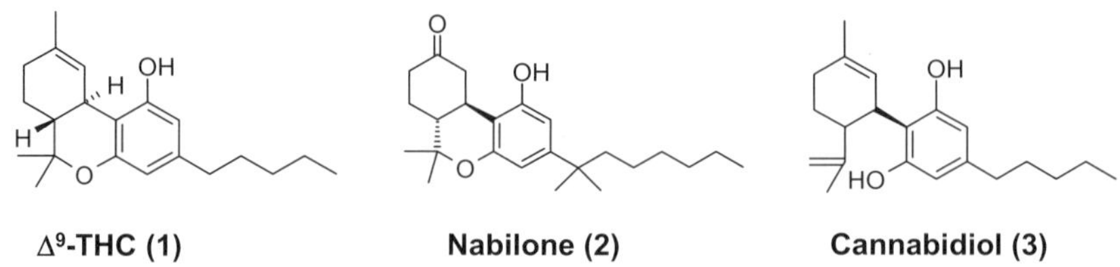

Figure 1. Chemical structures of Δ^9-THC **1**, nabilone **2**, and cannabidiol **3**.

Other potential therapeutic applications [7,8] of Δ^9-THC include several types of pain, multiple sclerosis, Gilles de la Tourette syndrome, epilepsy, certain forms of cancer [9], glaucoma and asthma.

Approximately 15 years ago, tremendous progress has been made in the molecular characterization of the endogenous cannabinoids and their receptors. First, the discovery and subsequent cloning of two distinct subtypes of cannabinoid receptors [10] (designated as CB_1 and CB_2) [11,12] opened new avenues to human CB receptor-based pharmacological screening. Their signal transduction pathways have been unraveled [13]. It is interesting to note that pharmacological evidence exists [14-16] for the presence of other cannabinoid receptor subtypes as well. Second, several endogenous substances (the so-called endocannabinoids) have been identified which act as agonists on the CB_1 and CB_2 receptors. Representative examples of these highly lipophilic, fatty acid-derived amides or esters are anandamide (**4**) and 2-arachidonoyl glycerol (**5**) (Figure 2).

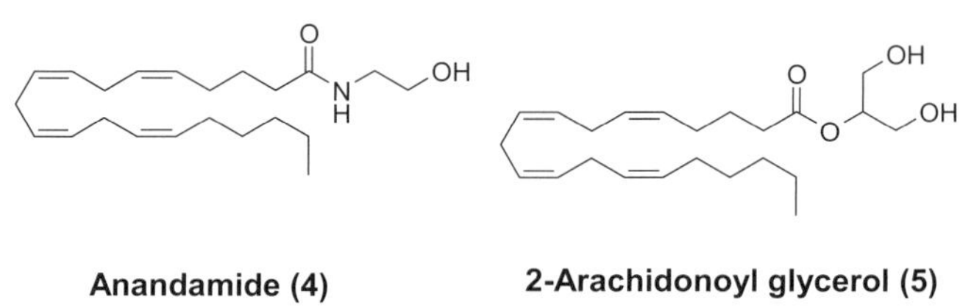

Figure 2. Chemical structures of two endocannabinoids, anandamide **4** and 2-arachidonoyl glycerol **5**.

These endocannabinoids, their putative transporters [17] and the endogenous enzymes which are involved in their formation and break-down (such as fatty amide hydrolase (FAAH) [18] together with the CB receptors constitute the endocannabinoid system [19-21]. It has been reported that cannabinoid signaling is altered [22,23] in neurological [24,25], psychiatric [26], cardiovascular, gastrointestinal, reproductive and eating disorders [27,28] and cancer [29].

The CB_1 receptor is abundantly expressed in several brain areas including basal ganglia, cerebellum, hippocampus and cortex as well as in peripheral tissues such as testis, eye, urinary bladder, ileum, liver, lung and adipocytes. Cannabinoids modulate neurotransmitter release via activation of presynaptic CB_1 receptors throughout the brain. The CB_2 cannabinoid receptor is predominantly found [30] in the immune system (spleen, tonsils, thymus), but more recently has also been detected [31,32] to some extend in peripheral nerve endings, microglial cells, astrocytes, cerebellum and brainstem. The homology between the CB_1 and CB_2 receptor [33] is relatively low (48% overall amino acid sequence identity and 68% sequence identity in the transmembrane region) which provides in theory the opportunity to design receptor subtype selective ligands. Hitherto, there have no agonists been reported which exhibit high selectivities on human CB_1 *versus* CB_2 receptors. However, the situation for selective CB_1 receptor antagonist design is clearly different which will become clear in the remainder of this book chapter. Interestingly, selective CB_2 receptor agonists and antagonists have also been discovered [34] albeit such compounds lie outside the scope of this chapter. Emphasis will herein be given to Solvay's research achievements in the fast-moving field of CB_1 receptor antagonists in relation with the current general state of the art, which has been reviewed [35-43] extensively. Numerous patents covering the field of CB_1 receptor antagonists have been published since 1994. Although the more general term "antagonist" is used in this chapter it is with the understanding that many CB_1 receptor antagonists behave as inverse agonists on the constitutively active receptor *in vitro* [44].

Therapeutic Prospects

Since the cannabinoid system plays a key role in human physiology the prospects for therapeutic applications of CB_1 receptor antagonists are promising. In 1998, the group of Colombo demonstrated [45] appetite suppression and weight loss in adult, non-obese Wistar rats after the administration of the prototypic, CB_1/CB_2 subtype selective CB_1 receptor antagonist rimonabant (SR141716) **6**. This finding gave enormous impetus to the CB_1 receptor antagonist research area [46] since the clinical need for an effective weight-reducing agent is rapidly increasing [47,48]. Studies of **6** have demonstrated efficacy in clinical obesity treatment. Moreover, rimonabant beneficially modulated obesity-associated metabolic parameters such as triglyceride levels and HDL cholesterol [49,50]. It has become apparent that CB_1 receptor antagonists like **6** act both through orexigenic effects and peripherally-mediated metabolic effects. Various tissues are involved [51] herein, including adipose tissue, liver [52], gastrointestinal tract, pancreas and skeletal muscle. Although rimonabant has recently been registered in Europe for the treatment of obesity, its approval in the USA by the FDA is still pending. Side effects such as dizziness, nausea, diarrhea, joint pain, anxiety and depression have been reported during clinical studies [49] of **6**. Rimonabant had weak anxiolytic-like activity in the elevated-plus maze and produced clear anxiolytic effects in the Vogel conflict test in rodents. Rimonabant was also reported to elicit anti-depressant-like effects in the forced-swimming test which were comparable [53] to those of the reference anti-depressant fluoxetine. These findings point to a role for the endocannabinoid system [21] in the modulation of emotional processes and unavoidable stress stimuli. Cannabinoid CB_1 receptor modulation and density [54] as well as the turnover rate of the endocannabinoid (**4**), and alterations in responsiveness of serotonin receptors following long-term cannabinoid treatment seem to be factors involved in depression [55], although their precise impacts require further clarification [56,57].

Rimonabant has been found active in many preclinical models predictive for addiction and drug dependence wherein it suppressed the reinforcing and rewarding properties of cocaine, heroin, alcohol and nicotine [58-60]. The endocannabinoid system is involved in the common neurobiological mechanism underlying drug addiction through the release of endocannabinoids in the ventral tegmental area in the brain [61].

However, the initial clinical results of **6** in a smoking cessation study [62] were not impressive.

CB_1 receptor antagonists may also have potential for the treatment of cognitive disorders. Albeit no clinical data are available, preclinical data revealed [63,64] that rimonabant antagonized memory impairment effects caused by Δ^9-THC and that **6** improved memory in an eight-arm radial maze task [65]. Moreover, in *in vivo* microdialysis studies **6** increased acetylcholine efflux in the rat medial prefrontal cortex [66].

Recently, it was shown that in addition to CB_2 receptor activation, antagonism of the CB_1 receptor holds promise for the treatment of liver fibrosis. CB_1 receptors were shown to be induced in human cirrhotic samples and in liver fibrogenic cells. Rimonabant inhibited the progression of fibrosis in three models of chronic liver injury [67]. As delineated in the introduction of this chapter, the CB receptor agonist Δ^9-THC has been reported to inhibit the growth of certain tumor cells. Intriguingly, CB_1 receptor antagonists have been reported [68] to inhibit human breast cancer cell proliferation and exerted a significant antiproliferative action by reducing the volume of induced tumors *in vivo* in mice. The treatment of inflammation and arthritis as two new potential indications for CB_1 receptor antagonists was recently proposed by researchers from Pfizer, based on results of **6**, its structural analogue AM251 and one of their proprietary CB_1 receptor antagonists in rodent *in vivo* models of inflammation [69]. Obesity has been associated with chronic inflammatory responses, including arthritis in relation to abnormal cytokine production [70]. Researchers from Sanofi-Aventis studied [71] anti-nociceptive and anti-inflammatory effects of **6** in lean and diet-induced obese female rats with adjuvant-induced arthritis. It was concluded that **6** might have therapeutic potential in obesity-associated inflammatory diseases and arthritis-associated pain. Sanofi-Aventis also claimed the use of derivatives of **6** for the prevention and treatment of chronic bronchitis and chronic obstructive bronchopneumopathy. It is clear that CB_1 receptor antagonism may lead to many physiological effects [72,73] which is in line with the high density of CB_1 receptors in the brain and their presence in many peripheral organs.

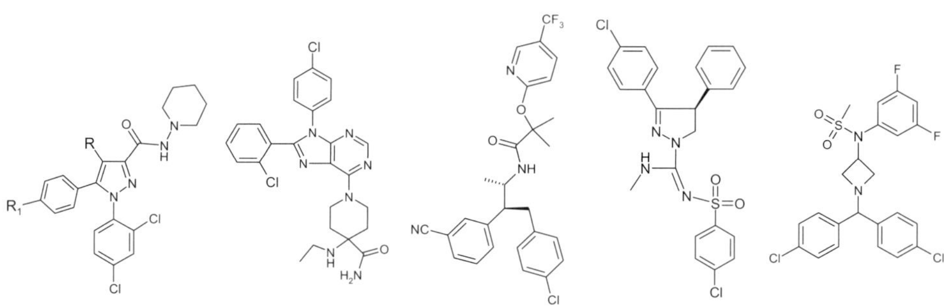

Rimonabant (6) R = CH_3, R_1 = Cl
Surinabant (6a) R = C_2H_5, R_1 = Br

Otenabant (7) **Taranabant (8)** **Ibipinabant (9)** **AVE-1625 (10)**

Figure 3. Chemical structures of CB_1 receptor antagonists in clinical development.

Rimonabant [74-76] was the first potent and selective CB_1 receptor antagonist discovered by Sanofi-Aventis which in the meantime has reached the European market for the treatment of obesity. Several selective CB_1 receptor antagonists are currently in clinical studies. The back-up compound [77] surinabant (**6a**) (SR147778) from Sanofi-Aventis is currently in Phase II trials for the treatment of smoking cessation, according to the Prous Science Integrity database. Pfizer is currently investigating otenabant (**7**) (CP-945,598) in a Phase III trial for the treatment of obesity. Merck & Co. are developing taranabant (**8**) (MK-0364) in Phase III clinical studies for the treatment of obesity [78-80]. Taranabant was also investigated in chronic cigarette smokers as an aid for smoking cessation. Solvay clinically co-develops ibipinabant (**9**) (SLV319) with Bristol-Myers Squibb [81]. AVE-1625 (**10**) is an azetidine derivative from Sanofi-Aventis which is currently undergoing Phase II clinical trials for the treatment of Alzheimer's type dementia, obesity and schizophrenia. AVE-1625 has also been studied for the treatment of insulin resistance.

Several other CB_1 receptor antagonists are in early clinical studies such as V-24343 (Vernalis, structure undisclosed) for the treatment of obesity and Org-50189 (Organon, structure undisclosed). Comparison of the chemical structures of the lipophilic compounds **6-10** reveals the general pharmacophore of these CB_1 receptor antagonists. Two adjacent aryl groups which may be substituted by one or two halogen atoms are linked via a heterocyclic or acyclic linker to a hydrogen acceptor moiety (usually a carbonyl or sulphonyl group). Furthermore, the compounds contain an additional lipophilic group which can be aliphatic or (hetero)aromatic [40,82]. The presence of a hydrogen acceptor moiety is related to its putative interaction with the Lys192 residue of the CB_1 receptor which is considered to stabilize the Lys192-Asp366 salt bridge of the intracellular end of transmembrane helices 3 and 6, thereby stabilizing the inactive R-state [83] of this constitutively active, G-protein coupled receptor. Based on these findings, Hein Coolen built an *in silico* receptor model of the inactive R-state of the human CB_1 receptor, thereby enabling structure-based design of cannabinoid CB_1 inverse agonists at Solvay.

Conformational Constraints in Cannabinoid CB_1 Receptor Antagonists

At the start of the cannabinoid program at Solvay under the supervision of Chris Kruse in 1996, only a few compounds including **6**, some aminoalkylindoles and the benzofuran derivative LY-320135 [84] were known as selective CB_1 receptor antagonists. It was decided to use **6** as a starting point for further design based on its high CB_1 receptor affinity, high CB_1/CB_2 receptor selectivity ratio and synthetic chemistry considerations.

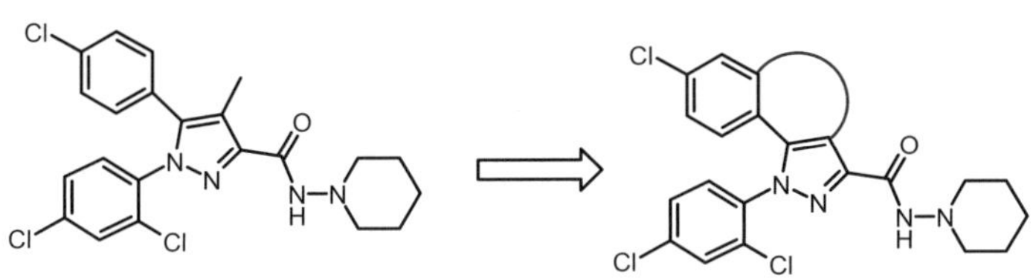

Rimonabant **11**

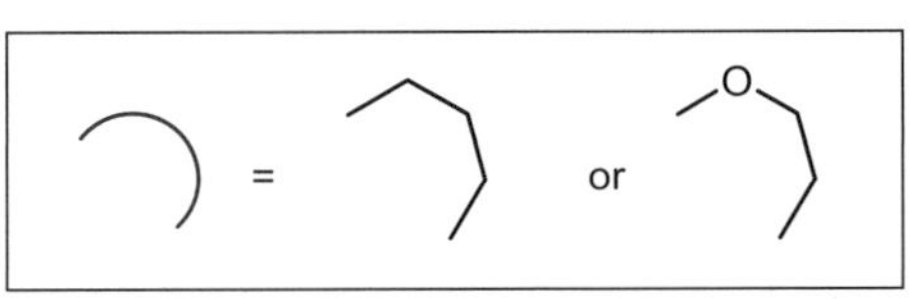

Figure 4. Conformational constraints used by Solvay, based on rimonabant.

An additional ring constraint from the ortho-position of its 5-aryl substituent to the 4-position on the pyrazole ring was envisioned to deliver rigidified analogs (**11**) of rimonabant, having a reduced number of rotatable bonds (Figure 4). In addition, such a conformational constraint would give valuable information on the bioactive conformation of **6** in the CB_1 receptor, thereby facilitating future cannabinoid drug design. The resulting compound **12** (Figure 5) was found active in an hCB_1 receptor binding assay as well as in a functional CB_1 receptor assay [85]. Structural comparison by molecular modeling analysis of **6** and **12** revealed that the difference in spatial conformation between both *in silico* minimized molecules is relatively small. However, compound **12** showed a poor bioavailability after oral administration which was attributed to its high lipophilicity (the calculated logP values (AlogP) of **6** and **12** amount to 6.6 and 7.2, respectively). In a later stage, compound **12** was published by an Italian group [86] coded as NESS 0327 and it also became apparent from the patent literature that Sanofi-Aventis had simultaneously been working along the same concept.

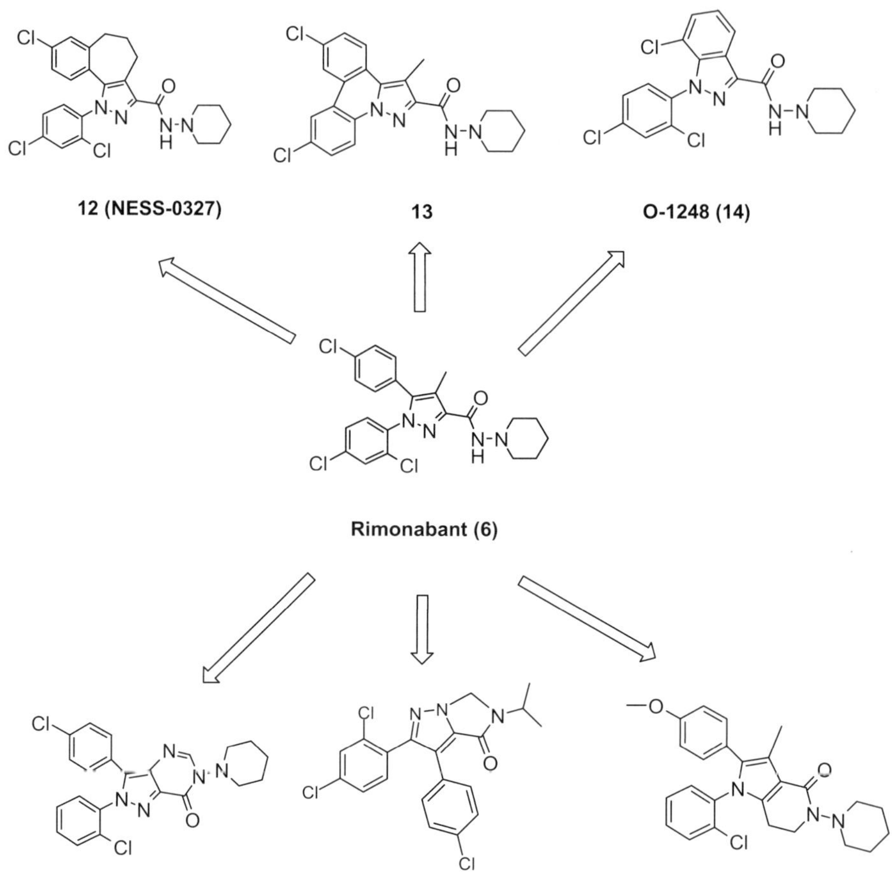

12 (NESS-0327) **13** **O-1248 (14)**

Rimonabant (6)

15 **16** **17**

Figure 5. Examples of rigidified analogs of rimonabant.

Other groups have also applied the concept of conformational constraint to produce rigidified analogs of rimonabant. Several of such conformational constraints have successfully been applied, differing in their ring sizes (five- to seven-membered rings), degree of saturation and attachment points to the pyrazole ring (either fused at its 1-5, 4-5 or 3-4 positions). Irradiation of **6** led [87] to the fused tetracyclic **13** which turned out to be a potent CB_1 receptor antagonist. The chloro-indazole derivative O-1248 (**14**) can be regarded [88] as a rigidified, but considerably less active analog of rimonabant. Carpino *et al* from Pfizer made many rimonabant-based rigidifications, resulting in potent CB_1 receptor antagonists like **15**, **16** and **17**. Although the *p*-methoxy substituted pyrrolopyridone (**17**) elicited a six-fold lower affinity for the human CB_1 receptor as compared to the corresponding *p*-chloro substituted derivative, compound (**17**) was found active in a Zucker rat obesity model after oral administration [89]. A close derivative of **17** was reported to bind analogously to the CB_1 receptor as **6**. However, it was postulated [90] that **15** presumably binds differently to the CB_1 receptor than rimonabant. It should be also noted that **16** has been reported to contain a hydrogen-bond acceptor group in another region of space than rimonabant, thereby assuming that the aromatic residues of both compounds are overlapping [43].

Bioisosterism

Bioisosteric replacement [91,92] forms a rational medicinal chemistry approach for the discovery of new leads, based on existing key ligands. In a broader context scaffold hopping [93,94] does the same, thereby producing compounds with altered molecular frameworks but with the preservation of key interaction points with the involved receptor. The three-dimensional structures of thiazoles, triazoles and imidazoles maintain a high similarity to that of the pyrazole. They have been applied at Solvay to discover pyrazole-based bioisosteric CB_1 receptor antagonists (Figure 6).

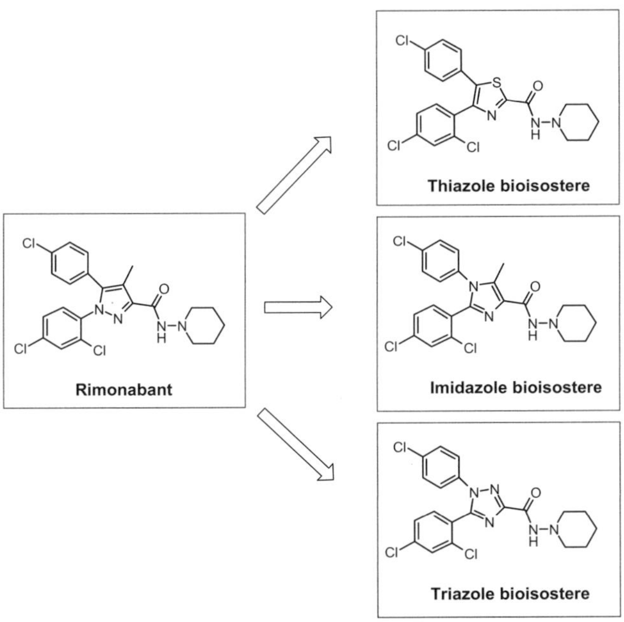

Figure 6. Bioisosteres based on rimonabant which were discovered by Solvay.

The imidazole bioisosteres showed a comparable profile as **6** in terms of their CB_1 receptor affinities, CB_1/CB_2 receptor subtype selectivity as well as their *in vivo* activities after oral administration in both a CB agonist-induced hypotension model and a CB agonist-induced hypothermia model [95]. In this 1,2-diarylimidazole class Solvay, Bayer and Merck & Co. have filed patent applications during a short time frame, followed by a publication by Neurocrine [96]. More recently, investigators from Bayer reported a detailed SAR analysis on 2-hydroxy-cyclohexylamides like **18** (Figure 7) which caused significant appetite suppression and dose-dependent reduction in body weight gain in rats after oral administration. The corresponding cyclohexylamide derivatives were very potent *in vitro* but showed poor pharmacokinetic properties. Presumably, the presence of the additional hydroxy-group in **18** enables intramolecular hydrogen-bonding, thereby improving its oral availability [97]. It is interesting to note that researchers from Merck & Co. discovered [98] 4,5-diarylimidazoles such as **19** as potent CB_1 receptor antagonists based on a hit (**20**) from high throughput screening, instead of pursuing a bioisosteric approach. Many more of such bioisosteric modifications have now been described [40,42] wherein the central pyrazole ring in rimonabant has been replaced by five-membered heteroaromatic rings (oxazoles, thiophenes, pyrroles, pyrazolines and imidazolines) or six-membered (hetero)aromatic rings (benzenes, pyrazines, pyridines and pyrimidines). Other structural classes such as (thio)hydantoins [99,100], 1,2,3,4-tetrahydroquinolines (Bristol-Myers Squibb) [36], 2,2-diaryl-1,3-benzodioxoles (Roche) [101] and benzofuran derivatives (Vernalis) [102] constitute less obvious scaffolds which are present in CB_1/CB_2 selective receptor antagonists [42].

The control of lipophilicity and its impact on ADME parameters and CNS penetrability was nicely demonstrated by appending a second ring to pyridine-derived CB_1 receptor antagonists. In this way the potent, but highly lipophilic CB_1 receptor antagonist **22** (ClogD = 7.9) - which was the result [103] of the hit optimization of **21** - could be converted into the naphthyridinone **23** (ClogD = 3.1). Whereas compound **23** showed an oral bioavailability of 93% and a brain/plasma ratio of 2.7 the much higher lipophilic **22** had a brain/plasma ratio less than 0.3, despite the fact that the total number of hydrogen-bond acceptors and donors in **23** is higher [104].

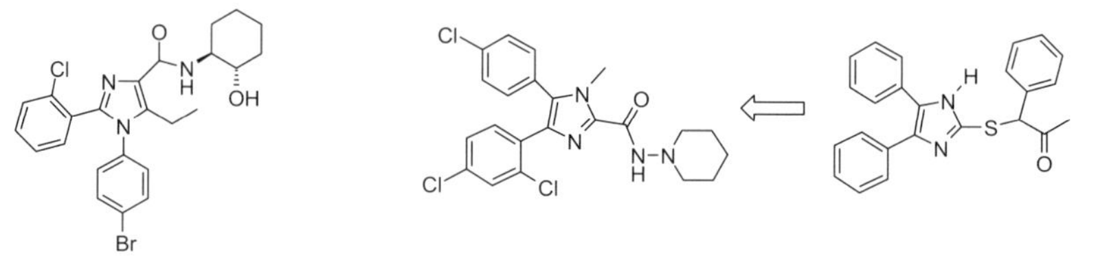

18: K_i hCB_1 = 3.7 nM   **19**: IC_{50} hCB_1 = 6.1 nM   **20**: IC_{50} hCB_1 = 7000 nM

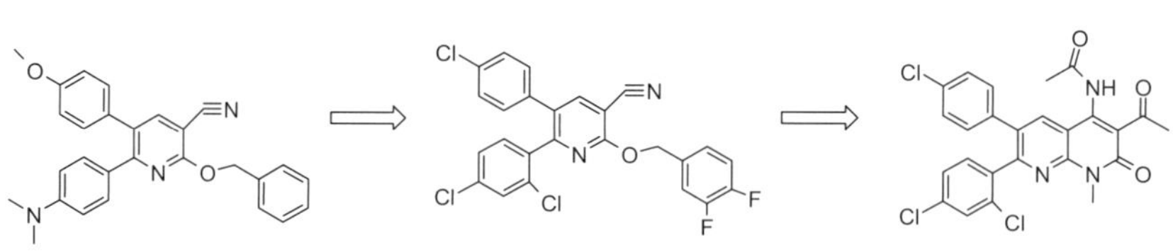

21: K_i hCB_1 = 530 nM   **22**: K_i hCB_1 = 1.3 nM   **23**: K_i hCB_1 = 7.5 nM

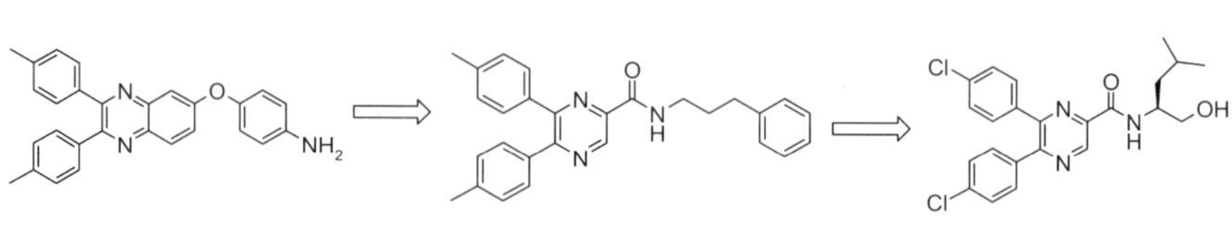

24: K_i hCB_1 = 650 nM   **25**: K_i hCB_1 = 18 nM   **26**: K_i hCB_1 = 14 nM

Figure 7. Optimizations of competitor CB_1 antagonists.

Although not stated by the authors, the high CNS penetrability of **23** may be attributed to intramolecular hydrogen-bonding effects in combination with its more compact structure. In line with this, Ellsworth *et al* (Bristol-Myers Squibb) reported that in their pyrazine series compound **25** - which was the result of optimization of the screening hit **24** - showed a significantly lower brain/plasma ratio as compared with **26** which was postulated to attain intramolecularly hydrogen-bonded conformations. In addition, **25** was inactive in a food intake model in food deprived rats after a 30 mg/kg oral dose, whereas **26** was found [105] active therein at a 10 mg/kg oral dose.

Researchers from AstraZeneca applied a scaffold hopping methodology steered by shape analogy with rimonabant to design 5,6-diarylpyrazine-2-carboxamides as potent CB_1 receptor antagonists [106].

CB_1 Screening at Solvay

A limited set of approximately 6000 compounds - including a stock diversity set of approximately 1500 compounds - was initially selected and screened in an hCB_1 receptor binding assay at Solvay. The outcome was DU513092 (**27**) which showed moderate affinity at the human CB_1 receptor (K_i = 196 nM) and a CB_1/CB_2 selectivity ratio greater than five (Figure 8). Compound **27** served as the starting point for further optimization [81], which led to its more potent (K_i = 16 nM) and CB_1/CB_2 selective (>60-fold) chloro-derivative (**28**). However, despite its nice *in vitro* profile **28** elicited low plasma levels after *per os* as well as *intraperitoneal* administration. Closer examination of its structure revealed a large polar moiety at the centre of the molecule. It was reasoned that substitution of the polar amidine moiety in **28** with small alkyl groups would give rise to **29** with a lower polar surface area (PSA). The PSA parameter has been shown to correlate well with drug transport properties in the gastrointestinal tract as well as with blood-brain barrier penetration [107,108]. The calculated PSA (cPSA) value of **29** was indeed much lower than the value for **28** but the effect of this structural modification on the human CB_1 receptor affinity of **29** (K_i = 280 nM) turned out to be detrimental. Nevertheless, the dimethyl-shielded **29** showed some activity after oral administration in a CB agonist-induced hypothermia model [81].

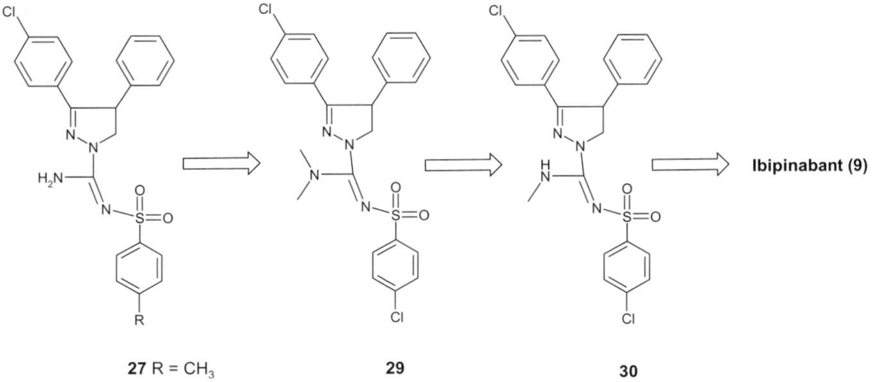

27 R = CH_3

29

30

28 R = Cl

Figure 8. Results from Solvay's CB_1 receptor antagonist screening program.

Therefore, it was decided to leave one methyl group in **29** to give **30**. The calculated PSA (cPSA) value of **30** was in between the values for **28** and **29** according to expectation but its affinity at the human CB_1 receptor was retained. Chiral separation of racemic **30** using preparative chiral HPLC, followed by crystallization and X-ray diffraction study elucidated the 4*S* configuration of ibipinabant (**9**). Bioanalytical studies revealed a CNS-plasma ratio of 1.7. A high CB_1/CB_2 receptor selectivity (~ 1000) was found in ibipinabant. The (4*R*)-distomer of ibipinabant showed ~ 100-fold lower CB_1 receptor affinity than **9**, indicating that this chiral ligand binds with a high degree of stereoselectivity to the CB_1 receptor. Molecular modeling studies showed a relatively close three-dimensional structural overlap between **9** and rimonabant (**6**). The docking result of **9** in the CB_1 receptor is depicted in Figure 9.

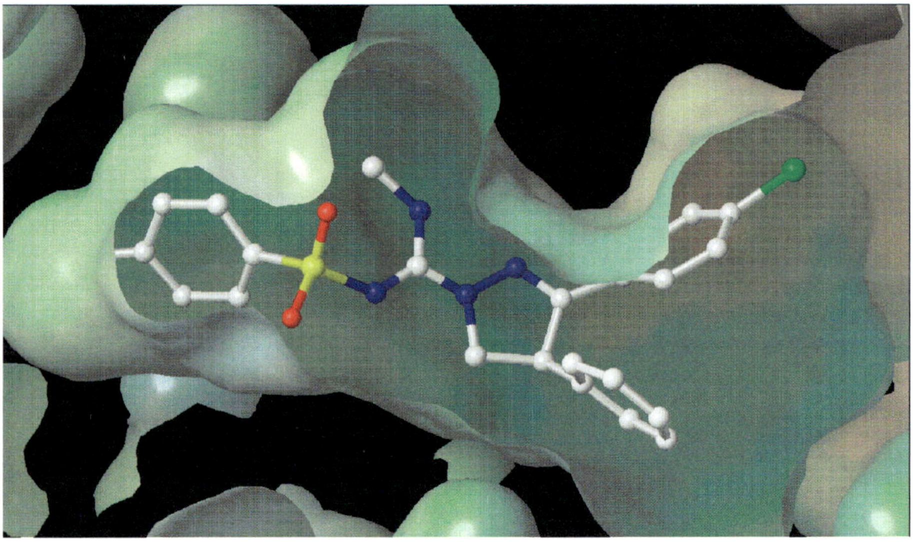

Figure 9. Representation of ibipinabant (**9**) docked in the cannabinoid CB_1 receptor.

Further analysis of the X-ray diffraction data of **9** revealed the presence of an intramolecular hydrogen bond which was confirmed by computational methods. Computational models and X-ray diffraction data indicated a different intramolecular hydrogen-bonding pattern in the *in vivo* inactive compound (**28**). In addition, X-ray diffraction studies of **28** revealed a tighter intermolecular packing than in **9** which also may contribute to its poorer absorption *in vivo*. A crystal lattice view of **9** is depicted in Figure 10.

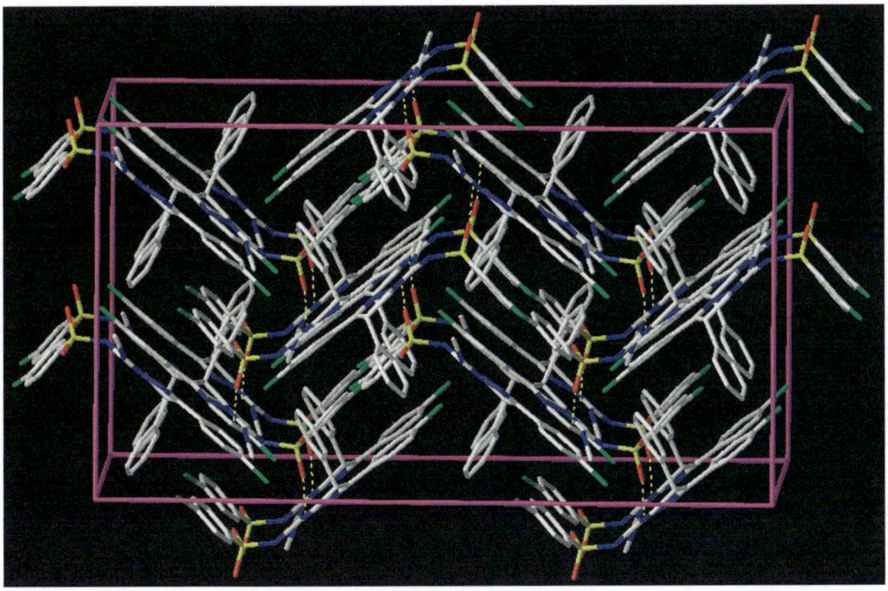

Figure 10. Crystal lattice view of ibipinabant (**9**).

It is clear that the methylation of the polar amidine moiety in the optimized hit (**28**) is pivotal in governing the oral bioavailability in the pyrazoline series, conceivably by subtly affecting the degree of dissolution rate in the gastrointestinal tract, as the result of a different intramolecular hydrogen-bonding pattern and crystal packing.

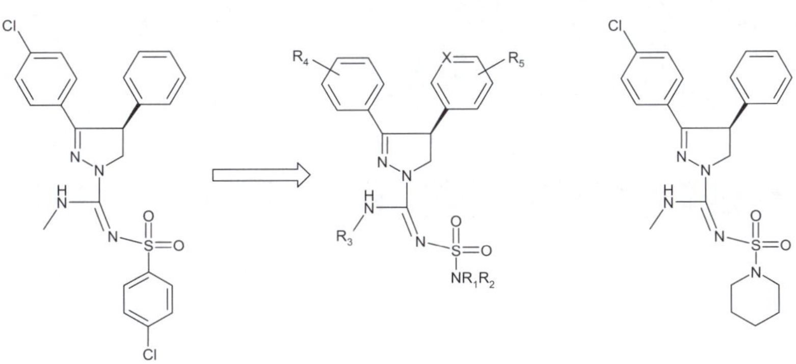

Ibipinabant (9) **31** **32**

Figure 11. Results from Solvay's CB_1 receptor antagonist back-up program.

In Solvay's CB_1 antagonist back-up program novel 3,4-diarylpyrazolines with lower lipophilicity than **9** were pursued (Figure 11). The key change is the replacement of the arylsulfonyl group in the original series by a dialkylaminosulfonyl moiety (Figure 9). Several compounds of general structure **31** in this series exhibit high CB_1 receptor affinities and high CB_1/CB_2 receptor subtype selectivities. Some representatives - such as the development candidate **32** - showed [109] pharmacological *in vivo* CB antagonistic activities after oral administration and are devoid of P-glycoprotein-mediated transport.

Recently, we reported [110] the discovery of novel CB_1 receptor antagonists, based on 1,5-diarylpyrazoline [111] and 1,2-diarylimidazoline templates.

Further Modifications of Rimonabant

The prototypic CB_1 receptor antagonist rimonabant has been used by many investigators as a structural template for further design of CB_1 receptor antagonists [36,40,42,43]. We envisaged that replacement of the 4-methyl substituent in rimonabant by a methylsulfanyl group would give rise to new chemical entities having retained cannabinoid receptor modulating properties, but significantly altered pharmacokinetic properties. Compound **33** indeed showed a high CB_1 receptor affinity (K_i = 10 nM), potent CB_1 functional activity (pA_2 = 8.6) and a high CB_1/CB_2 receptor selectivity. Surprisingly, the methylsulfanyl analogue (**33**) showed a higher *in vivo* activity (ED_{50} = 1.5 mg/kg) as compared with rimonabant (ED_{50} = 3.2 mg/kg) after oral administration in our rodent CB agonist-induced hypotension model. Due to the presence of the methylsulfanyl group in **33**, which was anticipated to be easily metabolized by the cytochrome P450 enzyme system [112] to the corresponding methylsulfinyl and methylsulfonyl group respectively, the metabolic pathway of compound **33** is expected to differ significantly from that of **6** (Figure 12).

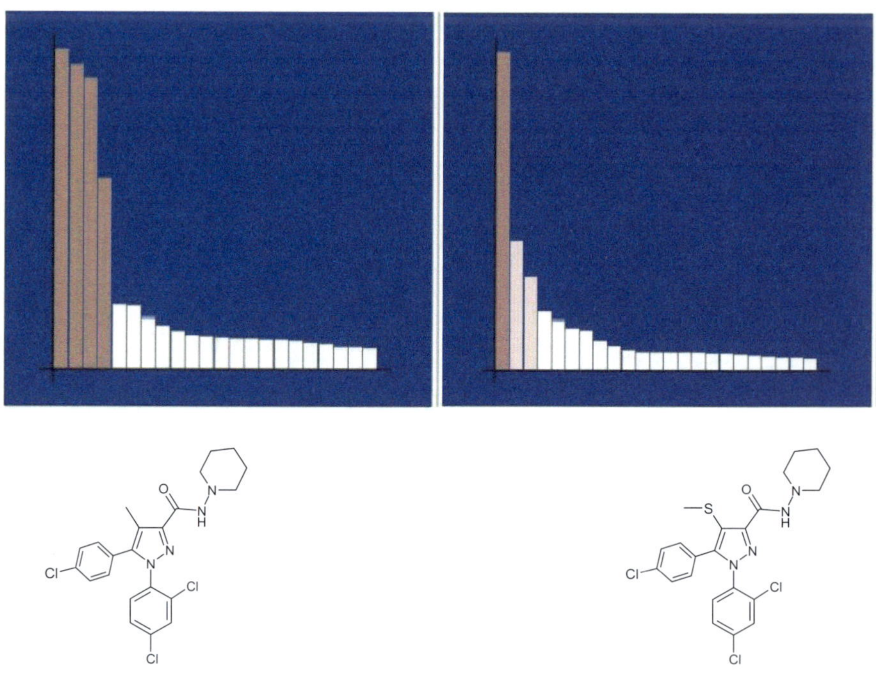

Rimonabant (6) **33**

Figure 12. MetaSite™ prediction of the site of metabolism of rimonabant (**6**) and its methylsulfanyl analogue (**33**).

Calculations [113] were carried out applying Cruciani's Metasite™ software [114] in order to predict *in silico* the primary sites of metabolism in phase I cytochrome-mediated reactions of the pyrazoles **6** and **33**, respectively. The results corroborate the hypothesis that the primary metabolic routes of the structurally related **6** and **33** might be different. Compound **33** is predicted by Metasite™ to metabolize primarily via oxidation of its sulphur atom (as represented by the left grey bar in the right graph in Figure 12) and to a much lesser extend on the 3- and 4-positions of its piperidine group, whereas **6** is predicted to be mainly oxidized on its 4-methyl group and the 3- and 4-positions of its piperidine moiety (as represented by the four left grey bars in the left graph in Figure 12). In conclusion, the replacement of the 4-methyl group in rimonabant by a methylsulfanyl group led to the novel compound (**33**) that exhibits potent CB_1 receptor antagonistic properties in combination with a different predicted biodegradation profile.

Summary

Well-known medicinal chemistry concepts such as the application of conformational constraints and scaffold hopping - in combination with results from pharmacological screening programs - have led to the discovery of selective CB_1 receptor antagonists from different chemical structure classes. The majority of these novel chemical entities can be described in terms of one general pharmacophore model.

References

[1] R. Mechoulam, L. Hanus, E. Fride. Towards cannabinoid drugs - revisited. *Prog. Med. Chem.* **35** (1998) 199-243.
[2] M.A. Elsohly and D. Slade. Chemical constituents of marijuana: the complex mixture of natural cannabinoids. *Life Sci.* **78** (2005) 539-548.
[3] Y. Gaoni and R. Mechoulam. Isolation, structure, and partial synthesis of an active constituent of hashish. *J. Am. Chem. Soc.* **86** (1964) 1646-1647.
[4] L.E. Long, D.T. Malone, D.A. Taylor. The pharmacological actions of cannabidiol. *Drugs Fut.* **30** (2005) 747-754.
[5] M.P. Barnes. Sativex®: clinical efficacy and tolerability in the treatment of symptoms of multiple sclerosis and neuropathic pain. *Expert Opin. Pharmacother.* **7** (2006) 607-615.
[6] J. Perez. Combined cannabinoid therapy via an oromucosal spray. *Drugs Today (Barc).* **42** (2006) 495-503.
[7] J.L. Croxford. Therapeutic potential of cannabinoids in CNS disease. *CNS Drugs* **17** (2003) 179-202.
[8] M. Ben Amar. Cannabinoids in medicine: a review of their therapeutic potential. *J. Ethnopharmacol.* **105** (2006) 1-25.
[9] I. Galve-Roperh, C. Sánchez, M.L. Cortés *et al.* Anti-tumoral action of cannabinoids: Involvement of sustained ceramide accumulation and extracellular signal-regulated kinase activation. *Nat. Med.* **6** (2000) 313-319. Comment in: *Nat. Med.* **6** (2000) 255-256.
[10] R.G. Pertwee (Eds). *Cannabinoid receptors*. Academic Press, London, UK (1995).
[11] L.A. Matsuda, S.J. Lolait, M.J. Brownstein *et al.* Structure of a cannabinoid receptor and functional expression of the cloned cDNA. *Nature* **346** (1990) 561-564.
[12] S. Munro, K.L. Thomas, M. Abu-Shaar. Molecular characterization of a peripheral receptor for cannabinoids. *Nature* **365** (1993) 61-65. Comment in: *Nature* **365** (1993) 12-13.
[13] A.C. Howlett, C.S. Breivogel, S.R. Childers *et al.* Cannabinoid physiology and pharmacology: 30 years of progress. *Neuropharmacology* **47** (2004) 345-358.
[14] J.L. Wiley and B.R. Martin. Cannabinoid pharmacology: implications for additional cannabinoid receptor subtypes. *Chem. Phys. Lipids* **121** (2002) 57-63.
[15] A.J. Brown. Novel cannabinoid receptors. *Br. J. Pharmacol.* **152** (2007) 567-575.
[16] E. Ryberg, N. Larsson, S. Sjögren *et al.* The orphan receptor GPR55 is a novel cannabinoid receptor. *Br. J. Pharmacol.* **152** (2007) 1092-1101.

[17] S.A. Moore, G.G. Nomikos, A.K. Dickason-Chesterfield *et al.* Identification of a high-affinity binding site involved in the transport of endocannabinoids. *Proc. Natl. Acad. Sci. U.S.A.* **102** (2005) 17852-17857. Comment in: *Proc. Natl. Acad. Sci. U.S.A.* **102** (2005) 17541-17542.
[18] M.K. McKinney and B.F. Cravatt. Structure and function of fatty acid amide hydrolase. *Annu. Rev. Biochem.* **74** (2005) 411-432.
[19] L. De Petrocellis, M.G. Cascio, V. Di Marzo. The endocannabinoid system: a general view and latest additions. *Br. J. Pharmacol.* **141** (2004) 765-774.
[20] D.M. Lambert and C.J. Fowler. The endocannabinoid system: drug targets, lead compounds, and potential therapeutic applications. *J. Med. Chem.* **48** (2005) 5059-5087.
[21] V. Di Marzo. Targeting the endocannabinoid system: to enhance or reduce? *Nat. Rev. Drug Discov.* **7** (2008) 438-455.
[22] V. Di Marzo, M. Bifulco, L. De Petrocellis. The endocannabinoid system and its therapeutic exploitation. *Nat. Rev. Drug Discov.* **3** (2004) 771-784.
[23] V. Di Marzo and S. Petrosino. Endocannabinoids and the regulation of their levels in health and disease. *Curr. Opin. Lipidol.* **18** (2007) 129-140.
[24] D. Centonze, A. Finazzi-Agrò, G. Bernardi *et al.* The endocannabinoid system in targeting inflammatory neurodegenerative diseases. *Trends Pharmacol. Sci.* **28** (2007) 180-187.
[25] V. Di Marzo. The endocannabinoid system for the development of new drugs for spasticity. *Drugs Fut.* **32** (2007) 341-352.
[26] B.L. Hungund, B.S. Basavarajappa, C. Vadasz *et al.* Ethanol, endocannabinoids, and the cannabinoidergic signaling system. *Alcohol. Clin. Exp. Res.* **26** (2002) 565-574.
[27] D. Cota, G. Marsicano, M. Tschöp *et al.* The endogenous cannabinoid system affects energy balance via central orexigenic drive and peripheral lipogenesis. *J. Clin. Invest.* **112** (2003) 423-431. Erratum in: *J. Clin. Invest.* **112** (2003) 323-326.
[28] S. Xie, M.A. Furjanic, J.J. Ferrara *et al.* The endocannabinoid system and rimonabant: a new drug with a novel mechanism of action involving cannabinoid CB_1 receptor antagonism – or inverse agonism – as potential obesity treatment and other therapeutic use. *J. Clin. Pharm. Ther.* **32** (2007) 209-231.
[29] M. Bifulco, C. Laezza, S. Pisanti *et al.* Cannabinoids and cancer: pros and cons of an antitumour strategy. *Br. J. Pharmacol.* **148** (2006) 123-135.
[30] T.W. Klein, C. Newton, K. Larsen *et al.* The cannabinoid system and immune modulation. *J. Leukoc. Biol.* **74** (2003) 486-496.
[31] M.D. van Sickle, M. Duncan, P.J. Kingsley *et al.* Identification and functional characterization of brainstem cannabinoid CB2 receptors. *Science* **310** (2005) 329-332.
[32] J.P. Gong, E.S. Onaivi, H. Ishiguro *et al.* Cannabinoid CB2 receptors: immunohistochemical localization in rat brain. *Brain Res.* **1071** (2006) 10-23.
[33] A.C. Howlett, F. Barth, T.I. Bonner *et al.* International Union of Pharmacology. XXVII. Classification of cannabinoid receptors. *Pharmacol. Rev.* **54** (2002) 161-202.
[34] K.H. Raitio, O.M. Salo, T. Nevalainen *et al.* Targeting the cannabinoid CB2 receptor: mutations, modeling and development of CB2 selective ligands. *Curr. Med. Chem.* **12** (2005) 1217-1237.
[35] J.H.M. Lange and C.G. Kruse. Recent advances in CB_1 cannabinoid receptor antagonists. *Curr. Opin. Drug Discov. Devel.* **7** (2004) 498-506.
[36] E.K. Högenauer. Latest advances in the cannabinoids. *Expert Opin. Ther. Pat.* **17** (2007) 1457-1476.
[37] R.A. Smith and Z. Fathi. Recent advances in the research and development of CB_1 antagonists. *IDrugs* **8** (2005) 53-66.
[38] L.W. Padgett. Recent developments in cannabinoid ligands. *Life Sci.* **77** (2005) 1767-1798.
[39] G.A. Thakur, S.P. Nikas, A. Makriyannis. CB1 cannabinoid receptor ligands. *Mini Rev. Med. Chem.* **5** (2005) 631-640.
[40] J.H.M. Lange and C.G. Kruse. Keynote review: medicinal chemistry strategies to CB_1 cannabinoid receptor antagonists. *Drug Discov. Today* **10** (2005) 693-702.
[41] F. Barth. CB1 cannabinoid receptor antagonists. *Annu. Rep. Med. Chem.* **40** (2005) 103-118.
[42] G.G. Muccioli and D.M. Lambert. Latest advances in cannabinoid receptor antagonists and inverse agonists. *Expert Opin. Ther. Pat.* **16** (2006) 1405-1423.
[43] D. Hepworth, P.A. Carpino, S.C. Black. Centrally acting anti-obesity agents. *Annu. Rep. Med. Chem.* **41** (2006) 77-97.
[44] R.G. Pertwee. Inverse agonism and neutral antagonism at cannabinoid CB_1 receptors. *Life Sci.* **76** (2005) 1307-1324.
[45] G. Colombo, R. Agabio, G. Diaz *et al.* Appetite suppression and weight loss after the cannabinoid antagonist SR 141716. *Life Sci.* **63** (1998) PL113-PL117.
[46] S.C. Black. Cannabinoid receptor antagonists and obesity. *Curr. Opin. Investig. Drugs* **5** (2004) 389-394.
[47] P.T. James, N. Rigby, R. Leach. The obesity epidemic, metabolic syndrome and future prevention strategies. *Eur. J. Cardiovasc. Prev. Rehabil.* **11** (2004) 3-8.

[48] D. Cooke and S. Bloom. The obesity pipeline: current strategies in the development of anti-obesity drugs. *Nat. Rev. Drug Discov.* **5** (2006) 919-931.
[49] L.F. van Gaal, A.M. Rissanen, A.J. Scheen *et al.* Effects of the cannabinoid-1 receptor blocker rimonabant on weight reduction and cardiovascular risk factors in overweight patients: 1-year experience from the RIO-Europe study. *Lancet* **365** (2005) 1389-1397.
[50] J. Antel, P.C. Gregory, U. Nordheim. CB_1 cannabinoid receptor antagonists for treatment of obesity and prevention of comorbid metabolic disorders. *J. Med. Chem.* **49** (2006) 4008-4016.
[51] M. Lafontan, P.V. Piazza, J. Girard. Effects of CB_1 antagonist on the control of metabolic functions in obese type 2 diabetic patients. *Diabetes Metab.* **33** (2007) 85-95.
[52] D. Osei-Hyiaman, M. DePetrillo, P. Pacher *et al.* Endocannabinoid activation at hepatic CB_1 receptors stimulates fatty acid synthesis and contributes to diet-induced obesity. *J. Clin. Invest.* **115** (2005) 1298-1305. Comment in: *J. Clin. Invest.* **115** (2005) 1130-1133.
[53] G. Griebel, J. Stemmelin, B. Scatton. Effects of the cannabinoid CB1 receptor antagonist rimonabant in models of emotional reactivity in rodents. *Biol. Psychiatry* **57** (2005) 261-267.
[54] B.L. Hungund, K.Y. Vinod, S.A. Kassir *et al.* Upregulation of CB_1 receptors and agonist-stimulated [^{35}S]GTPγS binding in the prefrontal cortex of depressed suicide victims. *Mol. Psychiatry* **9** (2004) 184-190.
[55] G. Gobbi, F.R. Bambico, R. Mangieri *et al.* Antidepressant-like activity and modulation of brain monoaminergic transmission by blockade of anandamide hydrolysis. *Proc. Natl. Acad. Sci. U.S.A.* **102** (2005) 18620-18625. Erratum in: *Proc. Natl. Acad. Sci. U.S.A.* **103** (2006) 2465.
[56] J.M. Witkin, E. Tzavara, R.J. Davis *et al.* A therapeutic role for cannabinoid CB_1 receptor antagonists in major depressive disorders. *Trends Pharmacol. Sci.* **26** (2005) 609-617.
[57] K.Y. Vinod and B.L. Hungund. Role of the endocannabinoid system in depression and suicide. *Trends Pharmacol. Sci.* **27** (2006) 539-545.
[58] M. Arnone, J. Maruani, F. Chaperon *et al.* Selective inhibition of sucrose and ethanol intake by SR 141716, an antagonist of central cannabinoid (CB1) receptors. *Psychopharmacology (Berl).* **132** (1997) 104-106.
[59] M. Solinas, L.V. Panlilio, K. Antoniou *et al.* The cannabinoid CB1 antagonist N-piperidinyl-5-(4-chlorophenyl)-1-(2,4-dichlorophenyl)-4-methylpyrazole-3-carboxamide (SR-141716A) differentially alters the reinforcing effects of heroin under continuous reinforcement, fixed ratio, and progressive ratio schedules of drug self-administration in rats. *J. Pharmacol. Exp. Ther.* **306** (2003) 93-102.
[60] B. Le Foll and S.R. Goldberg. Cannabinoid CB_1 receptor antagonists as promising new medications for drug dependence. *J. Pharmacol. Exp. Ther.* **312** (2005) 875-883.
[61] R. Maldonado, O. Valverde, F. Berrendero. Involvement of the endocannabinoid system in drug addiction. *Trends Neurosci.* **29** (2006) 225-232.
[62] S.L. Cox. Rimonabant hydrochloride: an investigational agent for the management of cardiovascular risk factors. *Drugs Today (Barc).* **41** (2005) 499-508.
[63] J.P. Terranova, J.J. Storme, N. Lafon *et al.* Improvement of memory in rodents by the selective CB1 receptor antagonist, SR 141716. *Psychopharmacology (Berl).* **126** (1996) 165-172.
[64] C. Castellano, C. Rossi-Arnaud, V. Cestari *et al.* Cannabinoids and memory: animal studies. *Curr. Drug Targets CNS Neurol. Disord.* **2** (2003) 389-402.
[65] M.C. Wolff and J.D. Leander. SR141716A, a cannabinoid CB_1 receptor antagonist, improves memory in a delayed radial maze task. *Eur. J. Pharmacol.* **477** (2003) 213-217.
[66] E.T. Tzavara, R.J. Davis, K.W. Perry *et al.* The CB1 receptor antagonist SR141716A selectively increases monoaminergic neurotransmission in the medial prefrontal cortex: implications for therapeutic actions. *Br. J. Pharmacol.* **138** (2003) 544-553.
[67] F. Teixeira-Clerc, B. Julien, P. Grenard *et al.* CB1 cannabinoid receptor antagonism: a new strategy for the treatment of liver fibrosis. *Nat. Med.* **12** (2006) 671-676. Comment in: *Hepatology* **45** (2007) 543-544; *J. Hepatol.* **46** (2007) 180-182; *Nat. Med.* **12** (2006) 608-610.
[68] D. Sarnataro, S. Pisanti, A. Santoro *et al.* The cannabinoid CB1 receptor antagonist rimonabant (SR141716) inhibits human breast cancer cell proliferation through a lipid raft-mediated mechanism. *Mol. Pharmacol.* **70** (2006) 1298-1306.
[69] D.M. Lambert. CB_1 cannabinoid receptor antagonism for treating inflammation and arthritis. *Expert Opin. Ther. Pat.* **17** (2007) 1027-1031.
[70] B. Costa. Rimonabant: more than an anti-obesity drug? *Br. J. Pharmacol.* **150** (2007) 535-537.
[71] T. Croci and E. Zarini. Effect of the cannabinoid CB_1 receptor antagonist rimonabant on nociceptive responses and adjuvant-induced arthritis in obese and lean rats. *Br. J. Pharmacol.* **150** (2007) 559-566. Comment in: *Br. J. Pharmacol.* **150** (2007) 535-537.
[72] M. Bifulco, C. Grimaldi, P. Gazzerro *et al.* Rimonabant: just an antiobesity drug? Current evidence on its pleiotropic effects. *Mol. Pharmacol.* **71** (2007) 1445-1456.
[73] V.E. Mendizábal and E. Adler-Graschinsky. Cannabinoids as therapeutic agents in cardiovascular disease: a tale of passions and illusions. *Br. J. Pharmacol.* **151** (2007) 427-440.

[74] A.K. Dutta, H. Sard, W. Ryan *et al.* The synthesis and pharmacological evaluation of the cannabinoid antagonist SR141716A. *Med. Chem. Res.* **5** (1994) 54-62.
[75] R. Lan, Q. Liu, P. Fan *et al.* Structure-activity relationships of pyrazole derivatives as cannabinoid receptor antagonists. *J. Med. Chem.* **42** (1999) 769-776.
[76] E.M. Nakamura-Palacios, J.M. Moerschbaecher, L.A. Barker. The pharmacology of SR 141716A: a review. *CNS Drug Rev.* **5** (1999) 43-58.
[77] S.A. Doggrell. Will the new CB_1 cannabinoid receptor antagonist SR-147778 have advantages over rimonabant? *Expert Opin. Investig. Drugs* **14** (2005) 339-342.
[78] L.S. Lin, T.J. Lanza Jr, J.P. Jewell *et al.* Discovery of N-[(1S,2S)-3-(4-chlorophenyl)-2-(3-cyanophenyl)-1- methylpropyl]-2-methyl-2-{[5-(trifluoromethyl)pyridin-2-yl]oxy}propanamide (MK-0364), a novel, acyclic cannabinoid-1 receptor inverse agonist for the treatment of obesity. *J. Med. Chem.* **49** (2006) 7584-7587.
[79] T.M. Fong, X.M. Guan, D.J. Marsh *et al.* Antiobesity efficacy of a novel cannabinoid-1 receptor inverse agonist, N-[(1S,2S)-3-(4-chlorophenyl)-2-(3-cyanophenyl)-1-methylpropyl]-2-methyl-2-[[5-(trifluoromethyl)pyridin-2-yl]oxy]propanamide (MK-0364), in rodents. *J. Pharmacol. Exp. Ther.* **321** (2007) 1013-1022.
[80] P. Cole, N. Serradell, E. Rosa *et al.* Taranabant. Cannabinoid CB_1 inverse agonist, antiobesity drug. *Drugs Fut.* **33** (2008) 206-212.
[81] J.H.M. Lange, H.K.A.C. Coolen, H.H. van Stuivenberg *et al.* Synthesis, biological properties, and molecular modeling investigations of novel 3,4-diarylpyrazolines as potent and selective CB(1) cannabinoid receptor antagonists. *J. Med. Chem.* **47** (2004) 627-643.
[82] P.H. Reggio. Pharmacophores for ligand recognition and activation/inactivation of the cannabinoid receptors. *Curr. Pharm. Des.* **9** (2003) 1607-1633.
[83] D.P. Hurst, D.L. Lynch, J. Barnett-Norris *et al.* N-(piperidin-1-yl)-5-(4-chlorophenyl)-1-(2,4-dichlorophenyl)-4-methyl-1H-pyrazole-3-carboxamide (SR141716A) interaction with LYS 3.28(192) is crucial for its inverse agonism at the cannabinoid CB_1 receptor. *Mol. Pharmacol.* **62** (2002) 1274-1287.
[84] C.C. Felder, K.E. Joyce, E.M. Briley *et al.* LY320135, a novel cannabinoid CB_1 receptor antagonist, unmasks coupling of the CB1 receptor to stimulation of cAMP accumulation. *J. Pharmacol. Exp. Ther.* **284** (1998) 291-297.
[85] A.R. Stoit, J.H.M. Lange, A.P. den Hartog *et al.* Design, synthesis and biological activity of rigid cannabinoid CB_1 receptor antagonists. *Chem. Pharm. Bull. (Tokyo)* **50** (2002) 1109-1113.
[86] S. Ruiu, G.A. Pinna, G. Marchese *et al.* Synthesis and characterization of NESS 0327: a novel putative antagonist of the CB_1 cannabinoid receptor. *J. Pharmacol. Exp. Ther.* **306** (2003) 363-370.
[87] M.E.Y. Francisco, J.P. Burgess, C. George *et al.* Structure elucidation of a novel ring-constrained biaryl pyrazole CB_1 cannabinoid receptor antagonist. *Magn. Reson. Chem.* **41** (2003) 265-268.
[88] C.E. Bass, G. Griffin, M. Grier *et al.* SR-141716A-induced stimulation of locomotor activity. A structure-activity relationship study. *Pharmacol. Biochem. Behav.* **74** (2002) 31-40.
[89] R.A. Smith, Z. Fathi, S.E. Brown *et al.* Constrained analogs of CB-1 antagonists: 1,5,6,7-Tetrahydro-4H-pyrrolo[3,2-c]pyridine-4-one derivatives. *Bioorg. Med. Chem. Lett.* **17** (2007) 673-678.
[90] P.A. Carpino, D.A. Griffith, S. Sakya *et al.* New bicyclic cannabinoid receptor-1 (CB_1-R) antagonists. *Bioorg. Med. Chem. Lett.* **16** (2006) 731-736.
[91] C.W. Thornber. Isosterism and molecular modification in drug design. *Chem. Soc. Rev.* **8** (1979) 563-580.
[92] L. Moreira Lima and E.J. Barreiro. Bioisosterism: a useful strategy for molecular modification and drug design. *Curr. Med. Chem.* **12** (2005) 23-49.
[93] J.L. Jenkins, M. Glick. J.W. Davies. A 3D similarity method for scaffold hopping from known drugs or natural ligands to new chemotypes. *J. Med. Chem.* **47** (2004) 6144-6159.
[94] H. Zhao. Scaffold selection and scaffold hopping in lead generation: a medicinal chemistry perspective. *Drug Discov. Today* **12** (2007) 149-155.
[95] J.H.M. Lange, H.H. van Stuivenberg, H.K.A.C. Coolen *et al.* Bioisosteric replacements of the pyrazole moiety of rimonabant: synthesis, biological properties, and molecular modeling investigations of thiazoles, triazoles and imidazoles as potent and selective CB_1 cannabinoid receptor antagonists. *J. Med. Chem.* **48** (2005) 1823-1838.
[96] B. Dyck, V.S. Goodfellow, T. Phillips *et al.* Potent imidazole and triazole CB_1 receptor antagonists related to SR141716. *Bioorg. Med. Chem. Lett.* **14** (2004) 1151-1154.
[97] R.A. Smith, Z. Fathi, F. Achebe *et al.* Optimization of imidazole amide derivatives as cannabinoid-1 receptor antagonists for the treatment of obesity. *Bioorg. Med. Chem. Lett.* **17** (2007) 2706-2711.
[98] C.W. Plummer, P.E. Finke, S.G. Mills *et al.* Synthesis and activity of 4,5-diarylimidazoles as human CB1 receptor inverse agonists *Bioorg. Med. Chem. Lett.* **15** (2005) 1441-1446.

[99] F. Ooms, J. Wouters, O. Oscari *et al.* Exploration of the pharmacophore of 3-alkyl-5-arylimidazolidinediones as new CB(1) cannabinoid receptor ligands and potential antagonists: synthesis, lipophilicity, affinity, and molecular modeling. *J. Med. Chem.* **45** (2002) 1748-1756. Erratum in: *J. Med. Chem.* **45** (2002) 4799.

[100] G.G. Muccioli, J. Wouters, C. Charlier *et al.* Synthesis and activity of 1,3,5-triphenylimidazolidine-2,4-diones and 1,3,5-triphenyl-2-thioxoimidazolidin-4-ones: characterization of new CB1 cannabinoid receptor inverse agonists/antagonists. *J. Med. Chem.* **49** (2006) 872-882.

[101] L. Alig, J. Alsenz, M. Andjelkovic *et al.* Benzodioxoles: novel cannabinoid-1 receptor inverse agonists for the treatment of obesity. *J. Med. Chem.* **51** (2008) 2115-2127.

[102] N. Foloppe, N.H. Allen, C.H. Bentley *et al.* Discovery of a novel class of selective human CB_1 inverse agonists. *Bioorg. Med. Chem. Lett.* **18** (2008) 1199-1206.

[103] L.C. Meurer, P.E. Finke, S.G. Mills *et al.* Synthesis and SAR of 5,6-diarylpyridines as human CB1 inverse agonists. *Bioorg. Med. Chem. Lett.* **15** (2005) 645-651. Erratum in: *Bioorg. Med. Chem. Lett.* **15** (2005) 1755.

[104] J.S. Debenham, C.B. Madsen-Duggan, T.F. Walsh *et al.* Synthesis of functionalized 1,8-naphthyridinones and their evaluation as novel, orally active CB1 receptor inverse agonists. *Bioorg. Med. Chem. Lett.* **16** (2006) 681-685.

[105] B.A. Ellsworth, Y. Wang, Y. Zhu *et al.* Discovery of pyrazine carboxamide CB1 antagonists: the introduction of a hydroxyl group improves the pharmaceutical properties and in vivo efficacy of the series. *Bioorg. Med. Chem. Lett.* **17** (2007) 3978-3982.

[106] J. Boström, K. Berggren, T. Elebring *et al.* Scaffold hopping, synthesis and structure-activity relationships of 5,6-diaryl-pyrazine-2-amide derivatives: a novel series of CB1 receptor antagonists. *Bioorg. Med. Chem.* **15** (2007) 4077-4084.

[107] K. Palm, P. Stenberg, K. Luthman *et al.* Polar molecular surface properties predict the intestinal absorption of drugs in humans. *Pharm. Res.* **14** (1997) 568-571.

[108] J. Kelder, P.D. Grootenhuis, D.M. Bayada *et al.* Polar molecular surface as a dominating determinant for oral absorption and brain penetration of drugs. *Pharm. Res.* **16** (1999) 1514-1519.

[109] J.H. Lange, H.H. van Stuivenberg, W. Veerman *et al.* Novel 3,4-diarylpyrazolines as potent cannabinoid receptor antagonists with lower lipophilicity. *Bioorg. Med. Chem. Lett.* **15** (2005) 4794-4798.

[110] J.H. Lange and C.G. Kruse. Cannabinoid CB1 receptor antagonists in therapeutic and structural perspectives. *Chem. Rec.* **8** (2008) 156-168.

[111] B.K. Srivastava, A. Joharapurkar, S. Raval *et al.* Diaryl dihydropyrazole-3-carboxamides with significant in vivo antiobesity activity related to CB1 receptor antagonism: synthesis, biological evaluation, and molecular modeling in the homology model. *J. Med. Chem.* **50** (2007) 5951-5966.

[112] C.M. Brown, B. Reisfeld, A.N. Mayeno. Cytochromes P450: a structure-based summary of biotransformations using representative substrates. *Drug Metab. Rev.* **40** (2008) 1-100.

[113] The .SD files of the compounds **6** and **33** respectively, served as the chemical structure input in Metasite™, version 2.8.6.

[114] G. Cruciani, E. Carosati, B. De Boeck *et al.* MetaSite: understanding metabolism in human cytochromes from the perspective of the chemist. *J. Med. Chem.* **48** (2005) 6970-6979.

Towards Drugs of the Future
C.G. Kruse and H. Timmerman (Eds.)
IOS Press, 2008

Author Index